How to Write and Publish a Scientific Paper

Eighth Edition

An essential guide for succeeding in today's competitive environment, this book provides beginning scientists and experienced researchers with practical advice on writing about their work and getting published. This brand new, updated edition also includes a new chapter on editing one's own work, a section on publicizing and archiving one's paper and updates on authorship, including information on new authorship criteria, and on the author identification number ORCID. The book guides readers through the processes involved in writing for and publishing in scientific journals: from choosing a suitable journal, to writing each part of the paper, to submitting the paper and responding to peer review, through checking the proofs. It covers ethical issues in scientific publishing; explains rights and permissions; and discusses writing grant proposals, giving presentations and writing for general audiences.

BARBARA GASTEL is Professor of Veterinary Integrative Biosciences, Humanities in Medicine, and Biotechnology at Texas A&M University. She has received awards and recognitions from the American Medical Writers Association, the Board of Editors in the Life Sciences, the Council of Science Editors, and Sigma Xi: The Scientific Research Society.

ROBERT A. DAY is Professor Emeritus of English at the University of Delaware, where he taught graduate and undergraduate courses in scientific writing. He has directed the publishing program of the American Society for Microbiology and served as managing editor of the *Journal of Bacteriology*. He also has been president of the Society for Scholarly Publishing and chairman of the Coun

How to Write and Publish a Scientific Paper

Eighth Edition

Barbara Gastel
Texas A&M University

Robert A. Day
University of Delaware

CAMBRIDGE
UNIVERSITY PRESS

University Printing House, Cambridge CB2 8BS, United Kingdom

One Liberty Plaza, 20th Floor, New York, NY 10006, USA

477 Williamstown Road, Port Melbourne, VIC 3207, Australia

4843/24, 2nd Floor, Ansari Road, Daryaganj, Delhi – 110002, India

79 Anson Road, #06–04/06, Singapore 079906

Cambridge University Press is part of the University of Cambridge.

It furthers the University's mission by disseminating knowledge in the pursuit of education, learning, and research at the highest international levels of excellence.

www.cambridge.org
Information on this title: www.cambridge.org/9781316640432
DOI: 10.1017/9781108105293

Third edition published 1989
Fourth edition published 1995
Fifth edition published 1998
Sixth edition published 2006
Seventh edition paperback published 2012
Eighth edition paperback published 2017

A hardback edition of How to Write and Publish a Scientific Paper, 8th edition was published by ABC-CLIO, LLC. in 2016.

Printed in the United Kingdom by TJ International Ltd. Padstow Cornwall.

This edition is not for sale in North America including the United States and its possessions, Canada and Mexico together.

A catalogue record for this publication is available from the British Library.

ISBN 978-1-316-64043-2 Paperback

Contents

PART VI: CONFERENCE COMMUNICATIONS

PART VII: SCIENTIFIC STYLE

Preface

Criticism and testing are of the essence of our work. This means that science is a fundamentally social activity, which implies that it depends on good communication. In the practice of science we are aware of this, and that is why it is right for our journals to insist on clarity and intelligibility.

—Hermann Bondi

Good scientific writing is not a matter of life and death; it is much more serious than that.

The goal of scientific research is publication. Scientists, starting as graduate students or even earlier, are measured primarily not by their dexterity in laboratory manipulations, not by their innate knowledge of either broad or narrow scientific subjects, and certainly not by their wit or charm; they are measured and become known (or remain unknown) by their publications. On a practical level, a scientist typically needs publications to get a job, obtain funding to keep doing research in that job, and gain promotion. At some institutions, publications are needed to obtain a doctorate.

A scientific experiment, no matter how spectacular the results, is not completed until the results are published. In fact, the cornerstone of the philosophy of science is based on the fundamental assumption that original research *must* be published; only thus can new scientific knowledge be authenticated and then added to the existing database that we call scientific knowledge.

It is not necessary for the plumber to write about pipes, nor is it necessary for the lawyer to write about cases (except *brief* writing), but the research scientist, perhaps uniquely among the trades and professions, must provide a document showing what he or she did, why it was done, how it was done, and what was learned from it. The key word is *reproducibility*. That is what makes science and scientific writing unique.

Thus, the scientist must not only "do" science but also "write" science. Bad writing can and often does prevent or delay the publication of good science.

Unfortunately, the education of scientists is often so overwhelmingly committed to the technical aspects of science that the communication arts are neglected or ignored. In short, many good scientists are poor writers. Certainly, many scientists do not like to write. As Charles Darwin said, "A naturalist's life would be a happy one if he had only to observe and never to write" (quoted by Trelease, 1958).

Most of today's scientists did not have a chance to take a formal course in scientific writing. As graduate students, they learned to imitate the style and approach of their professors and previous authors. Some scientists became good writers anyway. Many, however, learned only to imitate the writing of the authors before them—with all its defects—thus establishing a system of error in perpetuity.

The main purpose of this book is to help scientists and students of the sciences in all disciplines to prepare manuscripts that will have a high probability of being accepted for publication and of being completely understood when they are published. Because the requirements of journals vary widely from discipline to discipline, and even within the same discipline, it is not possible to offer recommendations that are universally acceptable. In this book, we present certain basic principles that are accepted in most disciplines.

Let us tell you a bit about the history of this book. The development of *How to Write and Publish a Scientific Paper* began many years ago, when one of us (Robert A. Day) taught a graduate seminar in scientific writing at the Institute of Microbiology at Rutgers University. It quickly became clear that graduate students in the sciences both wanted and needed *practical* information about writing. If a lecture was about the pros and cons of split infinitives, the students became somnolent; if it addressed how to organize data into a table, they were wide awake. Therefore, a straightforward "how to" approach was used for an article (Day 1975) based on the lecture notes. The article turned out to be surprisingly popular, and that led to the first edition of this book.

The first edition led naturally to the second edition and then to succeeding editions. Because this book is now being used in teaching programs in many colleges and universities, it seems especially desirable to keep it up to date. We thank those readers who kindly commented on previous editions, and we invite suggestions that may improve future editions. Please send suggestions and comments to Barbara Gastel at b-gastel@tamu.edu.

This edition, the eighth, is the third for which Barbara Gastel joins Robert A. Day—and the first for which Gastel is first author. Gastel remains grateful to Day for asking her to collaborate. We are delighted that our previous editions together have been translated into at least five languages, and we hope the current edition will be widely translated too.

In keeping with its title, this book has always focused primarily on writing and publishing scientific papers. It also has long provided broader advice on scientific communication. Beginning with the first edition, it has contained chapters to help readers write review papers, conference reports, and theses. Over time, chapters were added on other topics, such as how to present a paper orally and how to prepare a poster presentation. Recent editions also included chapters on approaching a writing project, preparing a grant proposal, writing about science in English as a foreign language, communicating science to the public, and providing peer review.

The current edition maintains this scope but has been substantially updated and otherwise revised. The electronic world of scientific communication has continued to evolve, and we have revised this book accordingly. Thus, for example, we now discuss using ORCID identifiers, avoiding predatory journals, and giving digital poster presentations. We have added a chapter on editing one's own work before submission, and we now include a section on publicizing and archiving one's paper after publication. The list of electronic resources has been expanded substantially. Cartoons have long been a popular feature of the book; we have retained favorites from previous editions and added several new cartoons by Jorge Cham (of PHD Comics), Sidney Harris, and others.

This book remains a "how-to book" or "cookbook," focusing mainly on points of practical importance. As in past editions, the book also contains some other items, such as cartoons and examples of humorous errors, intended to lighten the reading. Readers wishing to explore topics further are encouraged to consult works noted in the text or cited as references and to look at websites mentioned in this book.

Good scientific writing is indeed crucial. We hope this book will demystify writing and publishing a scientific paper and help you communicate about your work effectively, efficiently, and even enjoyably. Your success will be our greatest reward.

A Word to International Readers

For researchers throughout the world, communicating in English in standard Western formats has increasingly become the norm for sharing information widely. Thus, over the years *How to Write and Publish a Scientific Paper* has had many readers for whom English is not a native language. We hope the current edition will serve an even wider readership.

Aware of the diversity of our readers, we have tried especially hard in the current edition to present the main content in language easily understood by non-native speakers of English. One issue that we faced, however, was whether to retain the jokes that enlivened the book for many readers but sometimes confused readers from linguistic or cultural backgrounds other than our own. Because these jokes have been a distinctive feature of the book and one of its appeals, we have retained most of them in those chapters updated from early editions. However, because humor often does not translate well cross-culturally, we have limited its use in the more recently added chapters.

If, as an international reader, you occasionally encounter a silly-seeming story or comment in this book, do not worry that something is wrong or that you have missed an important point. Rather, realize that you are seeing some examples of American humor.

We welcome readers from throughout the world and hope they will find our book helpful in communicating science internationally. Suggestions for making the book more useful are appreciated at any time.

Acknowledgments

Over the years and over the editions, many colleagues and others have contributed directly or indirectly to this book. Those we have worked with in scientific publishing and academia have shared information and ideas. So have fellow members of the Council of Science Editors and the Society for Scholarly Publishing. Students and other users of the book have made suggestions. Many colleagues read and commented on manuscripts for early editions. Wura Aribisala, George Hale, Daniel Limonta Velázquez, Arkady Mak, Nancy Day Sakaduski, and Roberto Tuda Rivas read recent editions and offered thoughtful suggestions. Editors and production staff brought the work to publication. We thank all these people.

We also thank our families for their support, encouragement, and counsel. As preparations for this edition were beginning, life was ending for Sophie B. Gastel, mother of Barbara Gastel. It is to her memory that we dedicate this edition.

PART I

Some Preliminaries

CHAPTER 1

What Is Scientific Writing?

State your facts as simply as possible, even boldly. No one wants flowers of eloquence or literary ornaments in a research article.

—R. B. McKerrow

THE SCOPE OF SCIENTIFIC WRITING

The term *scientific writing* commonly denotes the reporting of original research in journals, through scientific papers in standard format. In its broader sense, scientific writing also includes communication about science through other types of journal articles, such as review papers summarizing and integrating previously published research. And in a still broader sense, it includes other types of professional communication by scientists—for example, grant proposals, oral presentations, and poster presentations. Related endeavors include writing about science for the public, sometimes called *science writing*.

THE NEED FOR CLARITY

The key characteristic of scientific writing is clarity. Successful scientific experimentation is the result of a clear mind attacking a clearly stated problem and producing clearly stated conclusions. Ideally, clarity should be a characteristic of any type of communication; however, when something is being said *for the first time,* clarity is essential. Most scientific papers, those published in our primary research journals, are accepted for publication precisely because they *do* contribute *new* knowledge. Hence, we should demand absolute clarity in scientific writing.

RECEIVING THE SIGNALS

Most people have no doubt heard this question: If a tree falls in the forest and there is no one there to hear it fall, does it make a sound? The correct answer is no. Sound is more than pressure waves, and indeed there can be no sound without a hearer.

And similarly, scientific communication is a two-way process. Just as a signal of any kind is useless unless it is perceived, a published scientific paper (signal) is useless unless it is both received *and* understood by its intended audience. Thus we can restate the axiom of science as follows: A scientific experiment is not complete until the results have been published *and understood.* Publication is no more than pressure waves unless the published paper is understood. Too many scientific papers fall silently in the woods.

UNDERSTANDING THE SIGNALS

Scientific writing is the transmission of a clear signal to a recipient. The words of the signal should be as clear, simple, and well-ordered as possible. In scientific writing, there is little need for ornamentation. Flowery literary embellishments—metaphors, similes, idiomatic expressions—are very likely to cause confusion and should seldom be used in research papers.

Science is simply too important to be communicated in anything other than words of certain meaning. And the meaning should be clear and certain not just to peers of the author, but also to students just embarking on their careers, to scientists reading outside their own narrow disciplines, and *especially* to those readers (most readers today) whose native language is other than English.

Many kinds of writing are designed for entertainment. Scientific writing has a different purpose: to communicate new scientific findings. Scientific writing should be as clear and simple as possible.

UNDERSTANDING THE CONTEXT

What is clear to a recipient depends both on what is transmitted and how the recipient interprets it. Therefore, communicating clearly requires awareness of what the recipient brings. What is the recipient's background? What is the recipient seeking? How does the recipient expect the writing to be organized?

Clarity in scientific writing requires attentiveness to such questions. As communication professionals advise, know your audience. Also know the conventions, and thus the expectations, for structuring the type of writing that you are doing.

ORGANIZATION AND LANGUAGE IN SCIENTIFIC WRITING

Effective organization is a key to communicating clearly and efficiently in science. Such organization includes following the standard format for a scientific paper. It also includes organizing ideas logically within that format.

In addition to organization, the second principal ingredient of a scientific paper should be appropriate language. This book keeps emphasizing proper use of English because many scientists have trouble in this area. All scientists must learn to use the English language with precision. A book (Day and Sakaduski 2011) wholly concerned with English for scientists is available.

If scientifically determined knowledge is at least as important as any other knowledge, it must be communicated effectively, clearly, in words of certain meaning. The scientist, to succeed in this endeavor, must therefore be literate. David B. Truman, when he was dean of Columbia University, said it well: "In the complexities of contemporary existence the specialist who is trained but uneducated, technically skilled but culturally incompetent, is a menace."

Given that the ultimate result of scientific research is publication, it is surprising that many scientists neglect the responsibilities involved. A scientist will spend months or years of hard work to secure data, and then unconcernedly let much of their value be lost because of a lack of interest in the communication process. The same scientist who will overcome tremendous obstacles to carry out a measurement to the fourth decimal place will be in deep slumber while a typographical error changes micrograms per milliliter to milligrams per milliliter.

English need not be difficult. In scientific writing, we say, "The best English is that which gives the sense in the fewest short words" (a dictum printed for some years in the *Journal of Bacteriology*'s instructions to authors). Literary devices, metaphors and the like, divert attention from substance to style. They should be used rarely in scientific writing.

CHAPTER 2

Historical Perspectives

History is the short trudge from Adam to atom.

—Leonard Louis Levinson

THE EARLY HISTORY

Human beings have been able to communicate for thousands of years. Yet scientific communication as we know it today is relatively new. The first journals were published about 350 years ago, and the *IMRAD* (introduction, methods, results, and discussion) organization of scientific papers has developed within about the past century.

Knowledge, scientific or otherwise, could not be effectively communicated until appropriate mechanisms of communication became available. Prehistoric people could communicate orally, of course, but each new generation started from essentially the same baseline because, without written records to refer to, knowledge was lost almost as rapidly as it was found.

Cave paintings and inscriptions carved onto rocks were among the first human attempts to leave records for succeeding generations. In a sense, today we are lucky that our early ancestors chose such media because some of these early "messages" have survived, whereas messages on less-durable materials would have been lost. (Perhaps many have been.) On the other hand, communication via such media was incredibly difficult. Think, for example, of the distributional problems the U.S. Postal Service would have today if the medium of correspondence were 100-lb (about 45-kg) rocks. It has enough troubles with 1-oz (about 28-g) letters.

The earliest book we know of is a Chaldean account of the Flood. This story was inscribed on a clay tablet in about 4000 BC, antedating Genesis by some 2,000 years (Tuchman 1980).

A medium of communication that was lightweight and portable was needed. The first successful medium was papyrus (sheets made from the papyrus plant and glued together to form a roll sometimes 20 to 40 ft [6–12 m] long, fastened to a wooden roller), which came into use about 2000 BC. In 190 BC, parchment (made from animal skins) came into use. The Greeks assembled large libraries in Ephesus and Pergamum (in what is now Turkey) and in Alexandria. According to Plutarch, the library in Pergamum contained 200,000 volumes in 40 BC (Tuchman 1980).

In AD 105, the Chinese invented paper, the dominant medium of written communication in modern times—at least until the Internet era. However, because there was no effective way of duplicating communications, scholarly knowledge could not be widely disseminated.

Perhaps the greatest single technical invention in the intellectual history of the human race was the printing press. Although movable type was invented in China in about AD 1100 (Tuchman 1980), the Western world gives credit to Johannes Gutenberg, who printed his 42-line-per-page Bible from movable type on a printing press in AD 1455. Gutenberg's invention was immediately and effectively put to use throughout Europe. By the year 1500, thousands of copies of hundreds of books were printed.

The first scientific journals appeared in 1665, when two journals, the *Journal des Sçavans* in France and the *Philosophical Transactions of the Royal Society of London* in England, began publication. Since then, journals have served as the primary means of communication in the sciences. As of 2014, there were nearly 35,000 peer-reviewed journals in science, technology, and medicine, of which more than 28,000 were in English. Altogether, these journals were publishing about 2.5 million articles per year (Ware and Mabe 2015, p. 6). The number of scientific papers published per year has been increasing exponentially (Bornmann and Mutz 2015).

THE ELECTRONIC ERA

When many older scientists began their careers, they wrote their papers in pen or pencil and then typed them on a typewriter or had a secretary do so. They or a scientific illustrator drew graphs by hand. They or a scientific photographer took photographs on film. They then carefully packaged a number of copies of the manuscript and sent them via postal service to a journal. The journal then mailed copies to the referees (peer reviewers) for evaluation, and the referees mailed them back with comments. The editor then mailed a decision letter to the scientist. If the paper was accepted, the scientist made the needed revisions and mailed back a final version of the manuscript. A copy editor edited the paper by hand, and a compositor re-keyboarded the manuscript. Once the paper was typeset, a copy was mailed to the scientist, who checked for typographical errors and mailed back corrections. Before the paper was published, the scientist ordered reprints of the paper, largely for fellow scientists who lacked access to libraries containing the journal or who lacked access to a photocopier.

Today the process has changed greatly. Word processors, graphics programs, digital photography, and the Internet have facilitated preparation and dissemination of scientific papers. Journals throughout the world have online systems for manuscript submission and peer review. Editors and authors communicate electronically. Manuscript editors typically edit papers online, and authors electronically receive typeset proofs of their papers for inspection. Journals are available online as well as in print—and sometimes instead of in print; increasingly, accepted papers become available individually online before appearing in journal issues. At some journals, electronic extras, such as appendixes and video clips, supplement online papers. Many journals are openly accessible online, either starting at the time of publication or after a lag period. In addition, readers often can access papers through the authors' websites or through resources at the authors' institutions, or the readers can request electronic reprints. Some of the changes have increased the technical demands on authors, but overall, the changes have hastened and eased the publication process and improved service to readers.

Whereas much regarding the mechanics of publication has changed, much else has stayed the same. Items that persist include the basic structure of a scientific paper, the basic process by which scientific papers are accepted for publication, the basic ethical norms in scientific publication, and the basic features of good scientific prose. In particular, in many fields of science, the IMRAD structure for scientific papers remains dominant.

THE IMRAD STORY

The early journals published papers that we call descriptive. Typically, a scientist would report, "First, I saw this, and then I saw that," or "First, I did this, and then I did that." Often the observations were in simple chronological order.

This descriptive style was appropriate for the kind of science then being reported. In fact, this straightforward style of reporting still is sometimes used in "letters" journals, case reports in medicine, geological surveys, and so forth.

By the second half of the nineteenth century, science was beginning to move fast and in increasingly sophisticated ways. Microbiology serves as an example.

Especially through the work of Louis Pasteur, who confirmed the germ theory of disease and developed pure-culture methods of studying micro-organisms, both science and the reporting of science made great advances.

At this time, methodology became all-important. To quiet his critics, many of whom were fanatic believers in the theory of spontaneous generation, Pasteur found it necessary to describe his experiments in exquisite detail. Because reasonably competent peers could reproduce Pasteur's experiments, the principle of *reproducibility of experiments* became a fundamental tenet of the philosophy of science, and a separate methods section led the way toward the highly structured IMRAD format.

The work of Pasteur was followed, in the early 1900s, by the work of Paul Ehrlich and, in the 1930s, by the work of Gerhard Domagk (sulfa drugs). World War II prompted the development of penicillin (first described by Alexander Fleming in 1929). Streptomycin was reported in 1944, and soon after World War II the mad but wonderful search for "miracle drugs" produced the tetracyclines and dozens of other effective antibiotics.

As these advances were pouring out of medical research laboratories after World War II, it was logical that investment in research would greatly increase. In the United States, this positive inducement to support science was soon (in 1957) joined by a negative factor when the Soviets flew *Sputnik* around our planet. In the following years, the U.S. government (and others) poured additional billions of dollars into scientific research.

Money produced science, and science produced papers. Mountains of them. The result was powerful pressure on the existing (and the many new) journals. Journal editors, in self-defense if for no other reason, began to demand that manuscripts be concisely written and well organized. Journal space became too precious to be wasted on verbosity or redundancy. The IMRAD format, which had been slowly progressing since the latter part of the nineteenth century, now came into almost universal use in research journals. Some editors espoused IMRAD because they became convinced that it was the simplest and most logical way to communicate research results. Other editors, perhaps not convinced by the simple logic of IMRAD, nonetheless hopped on the bandwagon because the rigidity of IMRAD did indeed save space (and expense) in the journals and because IMRAD made life easier for editors and referees by indexing the major parts of a manuscript.

The logic of IMRAD can be defined in question form: What question (problem) was studied? The answer is the introduction. How was the problem studied? The answer is the methods. What were the findings? The answer is the results. What do these findings mean? The answer is the discussion.

It now seems clear that the simple logic of IMRAD does help the author organize and write the manuscript, and IMRAD provides an easy road map for editors, referees, and ultimately readers to follow in reading the paper.

Although the IMRAD format is widely used, it is not the only format for scientific papers. For example, in some journals the methods section appears at the end of papers. In some journals, there is a combined results and discussion section. In some, a conclusions section appears at the end. In papers about research in which results of one experiment determine the approach taken in the next, methods sections and results sections can alternate. In some papers, especially in the social sciences, a long literature review section may appear near the beginning of the paper. Thus, although the IMRAD format is often the norm, other possibilities include IRDAM, IMRADC, IMRMRMRD, ILMRAD, and more.

Later in this book, we discuss components of a scientific paper in the order in which they appear in the IMRAD format. However, most of our advice on each component is relevant regardless of the structure used by the journal to which you will submit your paper. Before writing your paper, be sure, of course, to determine which structure is appropriate for the journal to which you will submit it. To do so, read the journal's instructions to authors and look at papers similar to yours that have appeared in the journal. These actions are parts of approaching a writing project—the subject of our next chapter.

CHAPTER 3

Approaching a Writing Project

Writing is easy. All you do is stare at a blank sheet of paper until drops of blood form on your forehead.

—Gene Fowler

ESTABLISHING THE MINDSET

The thought of preparing a piece of scientific writing can intimidate even the best writers. However, establishing a suitable mindset and taking an appropriate approach can make the task manageable. Perhaps most basic, remember that you are writing to communicate, not to impress. Readers of scientific papers want to know what you did, what you found, and what it means; they are not seeking great literary merit. If you do good research and present it clearly, you will please and satisfy readers. Indeed, in scientific writing, readers should notice mainly the content, not the style.

Realize that those reading your work want you to do well. They are not out to thwart you. Journal editors are delighted to receive good papers; ditto for the scientists they enlist as referees (peer reviewers) to help evaluate your work. Likewise, if you are a student, professors want you to do well. Yes, these people often make constructive criticisms. But they are not doing so because they dislike you; rather, they do so because they want your work to succeed. Do not be paralyzed by the prospect of criticism. Rather, feel fortunate that you will receive feedback that can help your writing to be its best.

PREPARING TO WRITE

In the laboratory, careful preparation helps experiments proceed smoothly and efficiently. Much the same is true of scientific writing. By preparing carefully before you start to compose a manuscript, you can make writing relatively easy and painless. Of course, in our unbiased view, preparing to write should include reading this book and keeping it on hand to consult. (Our publisher suggests buying a copy for your office or lab, a copy to use at home, and maybe one to keep in your car or boat.) But using this book is only a start. The following also can help.

Good writing is largely a matter of effective imitation. Therefore, obtain copies of highly regarded scientific papers in your research area, including papers in the journal to which you plan to submit your current work. Notice how these papers are written. For example: What sections do they include, and in what order? How long do the various sections tend to be? What types of subheadings, if any, tend to be included? How many figures and tables, and what types thereof, are typical? Especially if you are a non-native speaker of English, what seem to be some standard phrases that you could use in presenting your own work? Using published papers as models can prepare you to craft a manuscript that will be suitable to submit.

Successful writing also entails following instructions. Essentially every scientific journal issues instructions to authors. Following these instructions takes much of the guesswork out of writing and can save you from the unpleasant task of rewriting a paper because it did not meet the journal's specifications. If instructions are long (some journals' instructions run several pages or more), underline or highlight the key points to remember. Alternatively, you may list, on colored paper so you can easily find them, those points most relevant to the paper you will write. Also consider bookmarking on your computer the journal's instructions to authors, especially if the instructions encompass links for accessing different parts of their content.

For more detailed guidance—for instance, on nomenclature, reference formats, and grammar—instructions for authors often refer readers to standard style manuals. Among style manuals commonly used in the sciences are the following:

The ACS [American Chemical Society] Style Guide (Coghill and Garson 2006)
AMA [American Medical Association] Manual of Style (Iverson et al. 2007)
The Chicago Manual of Style (2010)
Publication Manual of the American Psychological Association (2010)
Scientific Style and Format (Style Manual Subcommittee, Council of Science Editors 2014)

New editions of these manuals come out from time to time. Increasingly, such manuals are available in online versions as well as in print. Look for the most recent edition of the style manual you will use. Commonly, you can find such style manuals in the reference sections of academic and other libraries. Many libraries also offer online access to style manuals. If you lack easy access, consider investing in the style manual(s) most commonly used in your research field. In any case, be ready to consult such manuals.

If you do not have reference-management software—for example, EndNote, Reference Manager, or RefWorks—now may be a good time to obtain it. Many universities make such software readily available and provide instruction in its use. Further information about such software appears in Chapter 15.

While you are gathering scientific content, ideas for your paper may occur to you. For example, you may think of a point to include in the discussion. Or you may come up with a good way to structure a table. Write down these ideas; consider creating for each section of your paper a file—either paper or electronic—in which to place them. Not only will recording your ideas keep them from escaping your memory, but having such ideas readily available to draw on can get your writing off to a quick start.

Once you have gathered and analyzed your data, speaking can be a fine transition to writing. If possible, present your work at a departmental seminar or local research day. Perhaps give an oral or poster presentation at a conference. Preparing to speak can help in formulating your article. Also, questions from listeners can help you to shape what you will write.

Research typically is a team endeavor. So is reporting on research. In the writing as in the research, different team members commonly take different roles. Sometimes one member drafts the whole paper and the others review and revise it. Other times, different members draft different parts of the paper and then circulate them for review. Whatever the case, clarify beforehand who will do what, and perhaps set a timetable. Maybe consider what software, if any, you will use to facilitate collaboration. Will you share drafts via Dropbox? Will you be using Google Docs? Will you use software designed specifically for academic collaboration? Discuss such matters before starting to write.

To facilitate writing, do lots of pre-writing. For example, stack copies of published papers in the order in which you plan to cite them. Make outlines. List points you wish to make in a given section, and sort and re-sort them until you are pleased with the order. Perhaps make a formal outline. By doing much of the thinking and organization beforehand, you can lower the activation energy needed to write a paper. In fact, such pre-writing can catalyze the writing process so well that you find yourself eager to write.

In preparing to write, realize that sometimes ideas must percolate for a while. If, for example, you cannot come up with an effective way to begin your paper or to structure a section, take a break. Exercise for a while, take a nap, or maybe discuss your work with someone. A solution may then occur to you.

DOING THE WRITING

Doing the writing means making time to do it. Most of us in science are busy. If writing must wait until we have extra time, it might never get done. Therefore, block out times to write. Indicate on your calendar or in your personal organizer the times that you have reserved for specific writing projects. Except in emergencies, do not let other tasks impinge on those times. Also, set deadlines. For example, promise yourself that you will draft a given section by Thursday. Or make clear to yourself that you will not leave for vacation until you have submitted a given item.

One highly published professor advocates the following approach (Zerubavel 1999): On a sheet of paper showing your weekly schedule hour by hour, cross out the times you are regularly unavailable—for example, times that you teach, have laboratory meetings, or have personal commitments. Then choose from the remaining times some to reserve for writing. In doing so, consider what times of day you tend to write most effectively. For example, if you are a night person, block out some evenings during which to write each week; perhaps save some morning time for more routine writing-related tasks, such as checking references. If you are a morning person, do the reverse.

When writing, you can start with whatever part of a manuscript you find easiest; there is no rule that you must write the introduction first. Many researchers like to begin by drafting the methods section, which tends to be the most straightforward to write. Many like to begin by drafting the figures and tables. Some like to start by drafting a preliminary reference list—or even the acknowledgments. And many authors leave until last the writing of the title and abstract. Once you have drafted one section, the momentum that you have established can facilitate writing the others. Feel free to draft the remaining sections in whatever order works best for you. Although the structure of Part II of this book parallels that of a scientific paper—with the first chapter addressing "How to Prepare the Title" and the last "How to Cite the References"—you can draft the parts of a scientific paper (and read these chapters) in whatever order works best for you.

Once you have established momentum, beware of dissipating it by interrupting your writing to search for small details. Rather, make notes to find the missing information; to identify them easily, write them in boldface type in your manuscript or use the "new comment" feature in Word. Also, if a manuscript will take more than one session to draft, consider how you can best maintain your momentum from session to session. Some authors like to stop in the

(© ScienceCartoonsPlus.com)

middle of a section while still going strong. Before ending their writing session, they jot down the next few points they wish to make. Thus, at their next writing session they can start quickly. Consider taking this approach.

Much like doing a piece of scientific research, crafting a scientific paper typically entails solving a series of problems in order to achieve the overall objective. In writing, as in research, often the problems have more than one reasonable solution, each with advantages and disadvantages. Yet writers sometimes worry that there is "one right way" (Becker 1986). Just how should a given item be worded? In just what format should a given illustration appear? How should a given part of the paper be organized? Often such questions have more than one good answer. Find one that seems reasonable and go with it. If it seems inadequate, or if a better solution occurs to you, you can make changes when you revise your manuscript.

REVISING YOUR WORK

Good writing tends to be largely a matter of good revising. No one will see your early drafts, and no one cares how rough they are (a comforting thought to those facing writer's block). The important thing is to revise your writing until it works well. First revise your writing yourself. Then show it to others and, using their feedback, revise your writing some more.

Revision is not just for students or other beginners. Researchers with long success in publishing revise the papers they write. After a presentation to a scientific-writing class, a well-known scientist and journal editor was asked, "Do you revise your work?" He answered: "If I'm lucky, only about 10 times."

In revising your work, ask yourself questions such as the following:

- Does the manuscript include all the information it should?
- Should any content be deleted?
- Is all the information accurate?
- Is all the reasoning sound?
- Is the content consistent throughout?
- Is everything logically organized?
- Is everything clearly worded?
- Have you stated your points briefly, simply, and directly? In other words, is everything concise?
- Are grammar, spelling, punctuation, and word use correct throughout?
- Are all figures and tables well designed?
- Does the manuscript comply with the instructions?

Information that can aid in answering some of these questions appears in later chapters of the book. For example, Chapters 10 through 13 describe the appropriate content and organization of the main sections of a scientific paper, and Chapters 30 through 34 address word usage and related subjects. In addition to reading these chapters before you write, consider consulting them as you revise your manuscript. Also, for further guidance, please see Chapter 41, which focuses mainly on editing one's own work.

Once your manuscript is nearly the best you can make it, show it to others and request their feedback. Years ago, scientists were advised, "Show your manuscript to a guy in your lab, a guy in a lab down the hall, and your wife." These days, such advice would rightly be viewed as inaccurate and sexist. Yet the concept remains valid. So, consider following this advice: Show your manuscript to an expert in your research specialty, who can help identify technical problems. Also show it to someone in your general field, who can note, for example, items that may be unclear to readers. And show it to an intelligent general reader—for instance, a friend in the humanities—who may identify problems that those interested mainly in the content tend to miss. In addition, consider also showing your manuscript to a professional scientific editor, as discussed in Chapter 41.

After receiving feedback from those reviewing your manuscript, consider how to apply it. Of course, follow those suggestions that you find useful. Even if a suggestion seems unsuitable, keep it in mind. Although you may disagree with it, it may alert you to a problem. For example, if a reader misinterpreted a point, you may try to state it more clearly. Comparing the various readers' comments may aid in this regard. If only one reader had difficulty with an item, you might dismiss it as a fluke. If, however, multiple readers did so, improvement probably is needed.

Revise your writing thoroughly. But avoid the temptation to keep revising it forever. No manuscript is perfect. Be satisfied with mere excellence. Journal editors and others will be pleased to receive the fine manuscripts you prepare by following the advice in this chapter and the rest of this book.

CHAPTER 4

What Is a Scientific Paper?

Without publication, science is dead.

—Gerard Piel

DEFINITION OF A SCIENTIFIC PAPER

A scientific paper is a written and published report describing original research results. That short definition must be qualified, however, by noting that a scientific paper must be written in a certain way, as defined by tradition, editorial practice, scientific ethics, and the interplay of printing and publishing procedures.

To properly define "scientific paper," we must define the mechanism that creates a scientific paper, namely, valid (that is, primary) publication. Abstracts, theses, conference reports, and many other types of literature are published, but such publications do not normally meet the test of valid publication. Further, even if a scientific paper meets all the other tests, it is not validly published if it is published in the wrong place. That is, a relatively poor research report, but one that meets the tests, is validly published if accepted and published in the right place (a primary journal or other primary publication); a superbly prepared research report is not validly published if published in the wrong place. Most of the government literature and conference literature, as well as institutional bulletins and other ephemeral publications, do not qualify as primary literature.

Many people have struggled with the definition of primary publication (valid publication), from which is derived the definition of a scientific paper. The Council of Biology Editors (CBE), now the Council of Science Editors (CSE), arrived at the following definition (Council of Biology Editors 1968, p. 2):

> An acceptable primary scientific publication must be the first disclosure containing sufficient information to enable peers (1) to assess observations, (2) to repeat experiments, and (3) to evaluate intellectual processes; moreover, it must be susceptible to sensory perception, essentially permanent, available to the scientific community without restriction, and available for regular screening by one or more of the major recognized secondary services (e.g., currently, Biological Abstracts, Chemical Abstracts, Index Medicus, Excerpta Medica, Bibliography of Agriculture, etc., in the United States and similar services in other countries).

At first reading, this definition may seem excessively complex, or at least verbose. But those who had a hand in drafting it weighed each word carefully and doubted that an acceptable definition could be provided in appreciably fewer words. Because it is important that students, authors, editors, and all others concerned understand what a scientific paper is and what it is not, it may be helpful to work through this definition to see what it really means.

"An acceptable primary scientific publication" must be "the first disclosure." Certainly, first disclosure of new research data often takes place via oral presentation at a scientific meeting. But the thrust of the CBE statement is that disclosure is more than disgorgement by the author; effective first disclosure is accomplished *only* when the disclosure takes a form that allows the peers of the author (either now or in the future) to fully comprehend and use that which is disclosed.

Thus, sufficient information must be presented so that potential users of the data can (1) assess observations, (2) repeat experiments, and (3) evaluate intellectual processes. (Are the author's conclusions justified by the data?) Then, the disclosure must be "susceptible to sensory perception." This may seem an awkward phrase, because in normal practice it simply means published; however, this definition provides for disclosure not just in terms of printed visual materials (printed journals and the no longer widely used media called microfilm and microfiche) but also in nonprint, nonvisual forms. For example, "publication" in the form of audio recordings, if that publication met the other tests provided in the definition, would constitute effective publication. And, certainly, electronic journals meet the definition of valid publication. What about material posted on a website? Views have varied and can depend on the nature of the material posted. For the most current information, consult materials from professional organizations and journals in your field.

Regardless of the form of publication, that form must be essentially permanent (often not the case for websites), must be made available to the scientific community without restriction (for example, in a journal that is openly accessible online or to which subscriptions are available), and must be made available to information-retrieval services (Biological Abstracts, Chemical Abstracts,

MEDLINE, etc.). Thus, publications such as newsletters, corporate publications, and controlled-circulation journals, many of which are of value for their news or other features, generally cannot serve as repositories for scientific knowledge.

To restate the CBE definition in simpler but not more accurate terms, primary publication is (1) the first publication of original research results, (2) in a form whereby peers of the author can repeat the experiments and test the conclusions, and (3) in a journal or other source document readily available within the scientific community. To understand this definition, however, we must add an important caveat. The part of the definition that refers to "peers of the author" is accepted as meaning prepublication peer review. Thus, by definition, scientific papers are published in peer-reviewed publications.

This question of definition has been belabored here for two reasons. First, the entire community of science has long labored with an inefficient, costly system of scientific communication precisely because it (authors, editors, and publishers) have been unable or unwilling to define primary publication. As a result, much of the literature has been buried in meeting abstracts, obscure conference reports, government documents, or books or journals of minuscule circulation. Other papers, in the same or slightly altered form, are published more than once; occasionally, this is due to the lack of definition as to which conference reports, books, and compilations are (or should be) primary publications and which are not. Redundancy and confusion result. Second, a scientific paper is, by definition, a particular kind of document containing specific kinds of information, typically in a prescribed (IMRAD) order. If the graduate student or the budding scientist (and even some of those scientists who have already published many papers) can fully grasp the significance of this definition, the writing task might be a great deal easier. Confusion results from an amorphous task. The easy task is the one in which you know exactly what must be done and in exactly what order it must be done.

ORGANIZATION OF A SCIENTIFIC PAPER

A scientific paper is organized to meet the needs of valid publication. It is, or should be, highly stylized, with distinctive and clearly evident component parts. The most common labeling of the component parts, in the basic sciences, is introduction, methods, results, and discussion (hence the acronym IMRAD). Actually, the heading "Materials and Methods" may be more common than the simpler "Methods," but the latter form was used in the acronym.

Some of us have taught and recommended the IMRAD approach for many years. The tendency toward uniformity has increased since the IMRAD system was prescribed as a standard by the American National Standards Institute, first in 1972 and again in 1979 (American National Standards Institute, 1979a). Some journals use a variation of IMRAD in which methods appear last rather than second. Perhaps we should call this IRDAM. In some journals, details regarding methods commonly appear in figure captions.

The basic IMRAD order is so eminently logical that, increasingly, it is used for many other types of expository writing. Whether one is writing an article about chemistry, archaeology, economics, or crime in the street, the IMRAD format is often the best choice.

This point is generally true for papers reporting laboratory studies and other experiments. There are, of course, exceptions. As examples, reports of field studies in the earth sciences and many clinical case reports in the medical sciences do not readily lend themselves to this kind of organization. However, even in these descriptive papers, the same logical progression from problem to solution is often appropriate.

Occasionally, the organization of laboratory papers must differ. If a number of methods were used to achieve directly related results, it might be desirable to combine the materials and methods and the results into an integrated experimental section. In some fields and for some types of results, a combined results and discussion section is usual or desirable. In addition, many primary journals publish notes or short communications, in which the IMRAD organization is modified.

Various types of organization are used in descriptive areas of science. To determine how to organize such papers and which general headings to use, refer to the instructions to authors of your target journal and look at analogous papers the journal has published. Also, you can obtain general information from appropriate source books. For example, types of medical papers are described by Huth (1999), Peat and others (2002), Taylor (2011), and contributors to a multiauthor guide (Hall 2013); types of engineering papers and reports are outlined by Michaelson (1990) and by Beer and McMurrey (2014). Indeed, even if a paper will appear in the IMRAD format, books on writing in one's own discipline can be worth consulting. Examples of such books include those in biomedical science by Zeiger (2000); the health sciences by Lang (2010); in chemistry by Ebel, Bliefert, and Russey (2004); and in psychology by Sternberg and Sternberg (2010).

In short, the preparation of a scientific paper has less to do with literary skill than with *organization*. A scientific paper is not literature. The preparer of a scientific paper is not an author in the literary sense. As an international colleague noted, this fact can comfort those writing scientific papers other than in their native language.

Some old-fashioned colleagues think that scientific papers should be literature, that the style and flair of an author should be clearly evident, and that variations in style encourage the interest of the reader. Scientists should indeed be interested in reading literature, and perhaps even in writing literature, but the communication of research results is a more prosaic procedure. As Booth (1981) put it, "Grandiloquence has no place in scientific writing."

Today, the average scientist, to keep up with a field, must examine the data reported in a very large number of papers. Also, English, the international language of science, is a second language for many scientists. Therefore, scientists (and of course editors) must demand a system of reporting data that is uniform, concise, and readily understandable.

SHAPE OF A SCIENTIFIC PAPER

Imagine that a friend visits your laboratory or office. The friend is unfamiliar with your research and wants to know about it. To orient your friend, first you identify your general research area and say why it is important. Then you state the specific focus of your research, summarize how you gathered your data, and say what you found. Finally you discuss the broader significance of your findings. The friend now has a new understanding—and, if you are lucky, he or she might buy you lunch.

Although intended for readers who are more knowledgeable, a scientific paper should take much the same approach: first providing broad orientation, then focusing narrowly on the specific research, and then considering the findings in wider context. Some have likened this shape for a scientific paper to an hourglass: broad, then narrow, then broad. Keeping this overall structure in mind can aid when writing individual parts of a paper and integrating them into a coherent whole.

OTHER DEFINITIONS

If *scientific paper* is the term for an original research report, how should this be distinguished from research reports that are not original, are not scientific, or somehow fail to qualify as scientific papers? Some specific terms are commonly used: *review paper, conference report,* and *meeting abstract.*

A review paper may review almost anything, most typically the recent work in a defined subject area or the work of a particular individual or group. Thus, the review paper is designed to summarize, analyze, evaluate, or synthesize information that *has already been published* (research reports in primary journals). Although much or all of the material in a review paper has previously been published, the problem of dual publication (duplicate publication of original data) does not normally arise because the review nature of the work is usually obvious—often from the title of the periodical, such as *Microbiology and Molecular Biology Reviews* or *Annual Review of Astronomy and Astrophysics.* Do not assume, however, that reviews contain nothing new. From the best review papers come new syntheses, new ideas and theories, and even new paradigms.

A conference report is a paper published in a book or journal as part of the proceedings of a symposium, national or international congress, workshop, roundtable, or the like. Such conferences commonly are not designed for the definitive presentation of original data, and the resultant proceedings (in a book or journal) do not qualify as primary publications. Conference presentations often are review papers, presenting reviews of the recent work of particular scientists or recent work in particular laboratories. Material at some conferences (especially the exciting ones) is in the form of preliminary reports, in which new, original data are presented, often accompanied by interesting speculation. But usually, these preliminary reports do not qualify, nor are they intended to qualify, as scientific papers. Later, often much later, such work may be validly published in a primary journal; by this time, the loose ends have been tied down, essential experimental details have been described (so that a competent worker could repeat the experiments), and previous speculation has matured into conclusions.

Therefore, the vast conference literature that appears normally is not *primary*. If original data are presented in such contributions, the data can and should be published (or republished) in an archival (primary) journal. Otherwise, the information may essentially be lost. If publication in a primary journal follows publication in a conference report, permission from the original publisher may be needed to reprint figures and other items (see Chapter 19, "Rights and Permissions"), but the more fundamental problem of dual publication normally does not and should not arise.

Meeting abstracts may be brief or relatively extensive. Although they can and generally do contain original information, they are not primary publications, and publication of an abstract should not preclude later publication of the full report.

Traditionally, there was little confusion regarding the typical one-paragraph abstracts published as part of the program or distributed along with the program at a national meeting or international congress. It was usually understood that many of the papers presented at these meetings would later be submitted for publication in primary journals. Sometimes conference organizers request extended abstracts (or *synoptics*). The extended abstract can supply almost as much information as a full paper; mainly it lacks the experimental detail. However, precisely because it lacks experimental detail, it cannot qualify as a scientific paper.

Those involved with publishing these materials should see the importance of careful definition of the different types of papers. More and more publishers, conference organizers, and individual scientists are agreeing on these basic definitions, and their general acceptance will greatly clarify both primary and secondary communication of scientific information.

CHAPTER 5

Ethics in Scientific Publishing

[A]ll scientists have an unwritten contract with their contemporaries and those whose work will follow to provide observations honestly obtained, recorded, and published.

—CBE Style Manual Committee

ETHICS AS A FOUNDATION

Before writing a scientific paper and submitting it to a journal—and indeed, before embarking on your research—you should know the basic ethical norms for scientific conduct and scientific publishing. Some of these norms may be obvious, others not. Therefore, a basic overview is provided below. Graduate students and others seeking further information on ethics in scientific publishing and more broadly in science may do well to consult *On Being a Scientist: Responsible Conduct in Research* (Committee on Science, Engineering, and Public Policy 2009), which contains both guidance and case studies and is accompanied online by a video. Other resources include ethics chapters in style manuals in the sciences.

AUTHENTICITY AND ACCURACY

That research reported in a journal should actually have been done may seem too obvious to mention. Yet cases exist in which the author simply made up data in a paper, without ever doing the research. Clearly, such "dry-labbing," or *fabrication,* is unethical. Fiction can be a grand pursuit, but it has no place in a scientific paper.

More subtle, and probably more common, are lesser or less definite deviations from accuracy: omitting outlying points from the data reported, preparing figures in ways that accentuate the findings misleadingly, or doing other tweaking. Where to draw the line between editing and distortion may not always be apparent. If in doubt, seek guidance from a more experienced scientist in your field—perhaps one who edits a journal.

The advent of digital imaging has given unethical researchers new ways to falsify findings. (Journal editors, though, have procedures to detect cases in which such falsification of images seems probable.) And ethical researchers may rightly wonder what manipulations of digital images are and are not valid. Sources of guidance in this regard include recent sets of guidelines for use and manipulation of scientific digital images (Cromey 2010, 2012).

For research that includes statistical analysis, reporting accurately includes using appropriate statistical procedures, not those that may distort the findings. If in doubt, obtain the collaboration of a statistician. Enlist the statistician early, while still planning the research, to help ensure that you collect appropriate data. Otherwise, ethical problems may include wasting resources and time. In the words of R.A. Fisher (1938), "To consult the statistician after an experiment is finished is often merely to ask him to conduct a *post mortem* examination."

ORIGINALITY

As discussed in the previous chapter, the findings in a scientific paper must be new. Except in rare and highly specialized circumstances, they cannot have appeared elsewhere in the primary literature. In the few instances in which republication of data may be acceptable—for example, in a more extensive case series or if a paper is republished in another language—the original article must be clearly cited, lest readers erroneously conclude that the old observations are new. To republish a paper (either in another language or for readers in another field) permission normally must be obtained from the journal that originally published the paper.

Beginning scientists sometimes wonder whether they may submit the same manuscript to two or more journals simultaneously. After all, a candidate can apply to several graduate programs at once and then choose among those offering acceptance. An analogous situation does not hold for scientific papers. Simultaneous submission wastes resources and is considered unethical. Therefore, begin by submitting your paper only to your first-choice journal. If that journal does not accept your paper, you can then proceed to the next journal on your list.

Originality also means avoiding "salami science" (or, for vegetarians, "cucumber science")—that is, thinly slicing the findings of a research project, as one might slice a sausage or cuke, in order to publish several papers instead of one (or, in the case of a large research project, many papers instead of a few). Good scientists respect the integrity of their research and do not divide it excessively for publication. Likewise, good hiring committees and promotion committees look at the content of publications, rather than only the number, and so are not fooled by salami science.

CREDIT

Good scientists build on each other's work. They do not, however, take credit for others' work.

If your paper includes information or ideas that are not your own, be sure to cite the source. Likewise, if you use others' wording, remember to place it in quotation marks (or to indent it, if the quoted material is long) and to provide a reference. Otherwise, you will be guilty of *plagiarism,* which the U.S. National Institutes of Health defines as "the appropriation of another person's ideas, processes, results, or words without giving appropriate credit" (National Institutes of Health 2010). To avoid inadvertent plagiarism, be sure to include information about the source when you copy or download materials others have written. To avoid the temptation to use others' wording excessively, consider drafting paragraphs without looking directly at the source materials; then look at the materials to check for accuracy.

In journal articles in most fields of science, it is unusual to include quotations from others' work. Rather, authors paraphrase what others have said. Doing so entails truly presenting the ideas in one's own way; changing a word or two does not constitute paraphrasing. On rare occasions—for example, when an author has phrased a concept extraordinarily well—quoting the author's own phrasing may be justified. If you are unsure whether to place in quotation marks a series of words from a publication, do so. If the quotation marks are unnecessary, an editor at the journal can easily remove them. If, however, they are missing but should have been included, the editor might not discover that fact (until, perhaps, a reader later does), or the editor might suspect the fact and send you an inquiry that requires a time-consuming search. Be cautious, and thus save yourself from embarrassment or extra work.

Resources to educate oneself about plagiarism, and thus learn better how to avoid it, include a tutorial from Indiana University (Frick and others 2016), an online guide to ethical writing (Roig 2003), and a variety of materials posted on websites of university writing centers. Another resource to consider is plagiarism-checking software. Such software helps identify passages of writing that seem suspiciously similar to text elsewhere; one can then see whether it does indeed appear to be plagiarized. Such software, such as Turnitin, is

"ASIDE FROM THE PLAGIARIZED PARTS, HOW IS IT?"

available at many academic institutions. Free plagiarism checkers, seemingly of varied quality, also exist. Many journal publishers screen submissions with plagiarism checking software, such as CrossCheck. Consider pre-screening your work yourself to detect and remove inadvertent plagiarism.

Also be sure to list as an author of your paper everyone who qualifies for authorship. (See Chapter 8 for more in this regard.) Remember as well to include in the acknowledgments those sources of help or other support that should be listed (see Chapter 14).

ETHICAL TREATMENT OF HUMANS AND ANIMALS

If your research involves human subjects or animals, the journal to which you submit your paper is likely to require documentation that they were treated ethically. Before beginning your study, obtain all needed permissions with regard to human or animal research. (In the United States, doing so entails having your research protocol reviewed by a designated committee at your institution.)

Then, in your paper, provide the needed statement(s) in this regard. For guidance, see the instructions to authors for the journal to which you are submitting your paper, and use as models papers similar to yours that have appeared in the journal. You may also find it useful to consult relevant sections of style manuals in the sciences. If in doubt, check with the publication office of the journal.

DISCLOSURE OF CONFLICTS OF INTEREST

Authors of scientific papers sometimes have *conflicts of interest*—that is, outside involvements that could, at least in theory, interfere with their objectivity in the research being reported. For example, they may own stock in the company making the product being studied, or they may be consultants to such a company.

Increasingly, it seems, journals are requiring authors to report such conflicts of interest. Some have checklists for doing so, and others ask more generally for disclosure. Journals vary in the degree to which they note conflicts of interest along with published papers (Clark 2005).

Ethics requires honest reporting of conflicts of interest. More importantly, ethics demands that such involvements not interfere with the objectivity of your research. Some scientists avoid all such involvements to prevent even the possibility of seeming biased.

CHAPTER 6

Where to Submit Your Manuscript

I've always been in the right place and time. Of course, I steered myself there.
—Bob Hope

WHY DECIDE EARLY, WHY DECIDE WELL

Too often, authors write scientific papers and then consider where to publish them. The decision, however, is best made early, before the writing begins. That way, the paper can be geared appropriately to the audience (for example, readers of a general scientific journal, a journal in your discipline as a whole, or a journal in your specialized research field). Also, thus you can initially prepare your manuscript in keeping with the journal's requirements, rather than having to revise it accordingly. Of course, if your first-choice journal does not accept your paper, you might need to revise your manuscript to suit another journal. But at least you will have avoided a round of revision.

In addition to deciding early on your first-choice journal, decide well. Choosing a journal carefully helps you to reach the most suitable audience, gain appropriate recognition, and avoid needless difficulties with publication. The decision where to submit the manuscript is important. Because of poor choices, some papers are delayed in publication, fail to receive sound review and revision, or lie buried in inappropriate journals. If you submit your manuscript to a poor choice of journal, one of three things can happen—all bad.

First, your manuscript may simply be returned to you, with the comment that your work "is not suitable for this journal." Often, however, this judgment is not made until *after* review of the manuscript. A "not suitable" notice after weeks or months of delay is not likely to make you happy.

(PEANUTS © 1982 Peanuts Worldwide LLC. Dist. By UNIVERSAL UCLICK. Reprinted with permission. All rights reserved.)

Second, if the journal is borderline in relation to your work, your manuscript may receive a poor or unfair review because the reviewers (and editors) of that journal may be only vaguely familiar with your specialty area. You may be subjected to the trauma of rejection even though the manuscript would be acceptable to the right journal. Or you could end up with a hassle over suggested revisions that you do not agree with and that do not improve your manuscript. And, if your manuscript really does have deficiencies, you would not be able to benefit from the sound criticism that would come from the editors of the right journal.

Third, even if your paper is accepted and published, your glee will be short-lived if you later find that your work is virtually unknown because it is buried in a publication that few in your intended audience read. Talking with colleagues can help prevent this situation.

Think about the appropriate readership. If, for example, you are reporting a fundamental study in physics, of course you should try to get your paper published in a prestigious international journal. On the other hand, suppose that your study relates to management of a disease found only in Latin America. In that situation, publication in *Nature* will not reach your audience—the audience that needs and can use your information. You should publish in an appropriate Latin American journal, probably in Spanish.

To start identifying journals to consider, recall what journals have published work similar to yours. The journals publishing the papers that you will cite are often journals to consider. Perhaps ask colleagues to suggest potential publication sites. To help determine whether a journal indeed seems to be a possibility, look in the journal or at its website for statements describing its purpose and scope. Look at some recent issues of the journal to see whether the journal publishes research such as yours and whether the papers are of the type you envision writing.

PRESTIGE AND IMPACT

If several journals seem suitable, does it matter which one you choose? Perhaps it shouldn't matter, but it does. There is the matter of *prestige*. It may be that progress in your career (job offers, promotions, grants, etc.) will be determined largely by the number of papers you publish. But not necessarily. It may well be that a wise old bird sitting on the faculty committee or the grant review panel will recognize and appreciate quality factors. A paper published in a "garbage" journal simply does not equal a paper published in a prestigious journal. In fact, the wise old bird (and there are quite a few of these in science) may be more impressed by the candidate with one or two solid publications in prestigious journals than by the candidate with 10 or more publications in second- or third-rate journals.

How do you tell the difference? It isn't easy, and of course there are many gradations. In general, however, you can form reasonable judgments by just a bit of bibliographic research. You will certainly know the important papers that have recently been published in your field. Make it your business to determine *where* they were published. If most of the real contributions in your field were published in Journal A, Journal B, and Journal C, you should probably limit your choices to those three journals. If Journals D, E, and F, upon inspection, contain only the lightweight papers, each could be eliminated as your first choice, even though the scope is right.

You may then choose among Journals A, B, and C. Suppose that Journal A is an attractive new journal published by a commercial publisher as a commercial venture, with no sponsorship by a society or other organization; Journal B is an old, well-known small journal published by a famous university, hospital, or museum; and Journal C is a large journal published by the principal scientific society in your field. In general (although there are many exceptions), Journal C (the society journal) is probably the most prestigious. It will also have the largest circulation (partly because of quality factors, and partly because society journals are less expensive than others, at least to society members). By publication in such a journal, your paper may have its best chance to make an impact on the community of scholars at whom you are aiming. Journal B might have almost equal prestige, but it might have a very limited circulation, which would be a minus; it might also be very difficult to get into, if most of its space is reserved for in-house material. Journal A (the commercial journal) might well have the disadvantage of low circulation (because of its comparatively high price, which is the result of both the profit aspect of the publisher and the lack of backing by a society or institution with a built-in subscription list). Publication in such a journal may result in a somewhat restricted distribution of your paper.

Be wary of new journals, especially those not sponsored by a society. (In particular, avoid *predatory journals*, which are discussed later in this chapter.) The circulation may be minuscule, and the journal might fail before it, and your paper, become known to the scientific world. Be wary of publishing in journals that are solely electronic unless you know that those evaluating your work for purposes such as promotion consider those journals as prestigious as journals with printed versions. On the other hand, be wary of publishing in the increasingly few journals that appear only in print, as scientists today expect important scientific literature to be accessible online.

One tool for estimating the relative prestige of journals in a given field is the electronic resource Journal Citation Reports, commonly available through academic libraries. With this resource, you can determine which journals have recently been cited most frequently, both in total and in terms of average number of citations per article published, or *impact factor* (Garfield 1999). Although not all good journals have impact factors computed, impact factor can be worth considering in judging the prominence of journals. If, in a given field, the average paper in Journal A is cited twice as frequently as the average paper in Journal B, it is likely that researchers find Journal A the more important journal. In some countries and institutions, impact factors of journals in which papers appear are among criteria considered when candidates are evaluated for promotion. However, limitations of the impact factor also should be noted. The impact factor indicates how much the papers in a journal are cited *on average*—not how much your paper will be cited if it appears in the journal. It does not indicate how much impact other than on citation the papers in a journal have—for example, how much they influence policy or clinical practice. And because different scientific fields have different citation practices, impact factors should not be used to compare importance of journals in different fields. For instance, in biochemistry and molecular biology, in which papers tend to cite many recent papers, the impact factor of the top-cited journal was 32.2 in the year 2014, but in geology it was 4.9. In short, although knowing a journal's impact-factor ranking in its field can help you assess the scientific importance of a journal, the impact factor does not say everything about the journal's quality and its suitability for your work. In journal selection as in much else in life, a multidimensional concept cannot validly be reduced to a single number.

Increasingly, experts have emphasized the need to include indicators other than impact factor when assessing the importance of a person's research. For example, the San Francisco Declaration on Research Assessment (2012), commonly called DORA, calls for using more varied approaches in evaluating research output. These approaches include—in addition to, most importantly, evaluating the scientific content of the article—using multiple journal-based metrics (rather than only impact factor) and looking at *article-level metrics*. Examples of the latter include how many times an article has been viewed,

"...BUT OUR MOST USEFUL PUBLICATION IS THE 'JOURNAL OF DON'T-DO-IT: IT'S-ALREADY-BEEN-DONE'."

(© ScienceCartoonsPlus.com)

downloaded, or bookmarked; how much attention it has received in social media and mass media; and how many times and where it has been cited (Tananbaum 2013). Noticing which journals' articles in your field tend to receive such attention can aid in identifying suitable journals for your papers.

ACCESS

Other items to consider when choosing journals can include *open access*—that is, the provision of articles online free of charge to all who may be interested. One consideration is whether to choose a journal (termed an *open-access journal*) that immediately provides open access to all its content. At such journals, which do not have subscriptions and so lack this source of income, the costs typically are defrayed at least in part by fees charged to authors. In some countries, these fees commonly are paid from grant funds; it can be wise to consider expected publication costs when preparing the budget for a grant. When authors, such as those in developing countries, cannot afford to pay the fees, the journal may waive or reduce them; if you cannot afford the normal publication fee for an open-access journal in which you hope to publish, contact the journal.

Access-related considerations for publishing in traditional journals can include whether to seek a journal for which the electronic version, initially available only to subscribers, becomes openly accessible relatively fast, for example, in a few months. Also, some journals give authors the option of making their articles freely accessible upon publication in return for paying a fee. Another consideration when publishing in a traditional journal is whether the journal allows rapid posting of articles on authors' or their institutions' websites. The website SHERPA/RoMEO (www.sherpa.ac.uk/romeo/) provides information about journals' policies in such regards.

AVOIDING PREDATORY JOURNALS

As noted, open-access journals typically charge authors fees as these journals lack income from subscriptions. Some dishonest people take advantage of this model by claiming to publish valid journals while instead just trying to get authors' money. These publishers of *predatory journals* may, for example, post all the papers that they receive, without peer review or editing. Or they might take authors' money and publish nothing. Submitting papers to such journals advances neither science nor one's career.

Such journals often market themselves vigorously, filling researchers' email with invitations to submit papers. How can you recognize, and thus avoid, predatory journals? Clues that a journal might be predatory include promises that seem too good to be true (for example, a guarantee to publish all submissions within a week), a website with many typographical and other errors, inclusion of what seem to be fake metrics (such as "impact index"), and lack of good articles (or any articles at all) on the journal's website. On the other hand, indications that a journal is likely to be valid include publication of good articles that you already have seen and inclusion of the journal in major bibliographic databases. If you think that a journal might be predatory, consider consulting Beall's List (scholarlyoa.com/publishers/), compiled by academic librarian Jeffrey Beall. This list of "potential, possible, or probable predatory scholarly open-access publishers" can aid in evaluating one's suspicions.

Especially if you are inexperienced in publishing, perhaps consult a mentor or senior colleague if you think a journal that you are considering might be predatory. In fact, in any case, such consultation can be wise before finalizing one's choice of a journal.

OTHER FACTORS TO CONSIDER

In choosing a journal, other factors also can merit consideration. One such factor is speed of publication. Increasingly, journals have been publishing papers online before they appear in print or are included in an online issue. You may find it worthwhile to check whether a journal publishes individual articles online first and, if so, how quickly it does so.

The time from acceptance to publication in a journal issue generally reflects the frequency of the journal. For example, the publication lag of a monthly journal is almost always shorter than that of a quarterly journal. Assuming equivalent review times, the additional delay of the quarterly will range up to 2 or 3 months. Since the publication lag, including the time of editorial review, of many (probably most) monthlies ranges between 4 and 7 months, the lag of the quarterly is likely to run up to 10 months. Remember, also, that many journals, whether monthly, bimonthly, or quarterly, have backlogs. It sometimes helps to ask colleagues what their experience has been with the journal(s) you are considering. If the journal publishes "received for publication" dates, you can figure out for yourself the average lag time.

Even in this electronic age, quality of printing can be a consideration. In biology, the journals published by the American Society for Microbiology and by the Rockefeller University Press traditionally have been especially noted for their high standards in this respect. Whatever your field, look at the reproduction quality of the journal if it will be important to you.

Finally, consider likelihood of acceptance. Clearly, not every paper is important enough and of broad enough interest to appear in *Science* or *Nature*. Rather, most papers belong in journals in their disciplines or subdisciplines.

Even within specific fields, some papers are of great enough importance for publication in first-line journals, whereas many others can better find homes elsewhere. In initially submitting your paper, aim high, generally for the broadest and most prestigious journal in which your paper seems to have a realistic chance of publication. To decide on this journal, perhaps look again at candidate journals and consult colleagues. Choosing a journal that is appropriate with regard to subject matter, audience, prestige, access, selectivity, and other factors can help ensure that your paper will be published without undue delay—and that it will be read and recognized by those it should reach.

USING INSTRUCTIONS TO AUTHORS

In considering where to submit your paper, you might have looked at some journals' instructions to authors to learn more about the journals' scopes, audiences, or requirements. If you have not yet obtained the instructions for the journal you chose, do so before starting to write. Typically, these instructions appear on the website of the journal. In addition, instructions from more than 6,000 biomedical journals can be accessed through the website Instructions to Authors in the Health Sciences, mulford.utoledo.edu/instr. This site also includes links to sets of guidelines that many medical journals follow.

If you do not find instructions to authors immediately, keep looking. Sometimes their location on the journal website is not initially apparent. Also, instructions to authors can have a variety of other names, such as information for authors, guide for authors, and submission instructions. If, after careful searching, you still do not find the instructions, consider asking a more experienced researcher or a librarian for help or contacting the office of the journal. Also, a lack of instructions can be a clue that a journal is predatory rather than legitimate.

Read the instructions for authors thoroughly before starting to prepare your paper. Among questions these instructions may answer are the following:

- Does the journal include more than one category of research article? If so, in what category would yours fit?
- What is the maximum length of articles? What is the maximum length of abstracts?
- Does the journal have a template for articles? If so, how can it be accessed?
- Does the journal post supplementary material online, if applicable? If so, how should this material be provided?
- What sections should the article include? What guidelines should be followed for each?
- What guidelines should be followed regarding writing style?

(www.phdcomics.com)

- How many figures and tables are allowed? What requirements does the journal have for figures and tables?
- In what format should references appear? Is there a maximum number of references?
- In what electronic format should the paper be prepared? Should figures and tables be inserted within the text, or should they appear at the end or be submitted as separate files? Is there an online submission system to use?

Underline, highlight, or otherwise note key points to remember. Then consult the instructions to authors as you prepare the paper. Following the instructions from the outset will save time overall.

Also look carefully at some recent issues of the journal. Pay particular attention to those aspects of editorial style that tend to vary widely from journal to journal. These aspects include the style of literature citation, the use of headings and subheadings, and the design of tables and figures.

Shortly before submitting your manuscript, check the instructions to authors again, and ensure they have been followed. If the instructions include a checklist, use it. By following the instructions carefully, you will facilitate publication of your manuscript from the time you begin to draft it.

PART II

Preparing the Text

CHAPTER 7

How to Prepare the Title

First impressions are strong impressions; a title ought therefore to be well studied, and to give, so far as its limits permit, a definite and concise indication of what is to come.

—T. Clifford Allbutt

IMPORTANCE OF THE TITLE

In preparing a title for a paper, you would do well to remember one salient fact: This title will be read by thousands of people. Perhaps few people, if any, will read the entire paper, but many people will read the title, either in the original journal, in one of the secondary (abstracting and indexing) databases, in a search engine's output, or otherwise. Therefore, all words in the title should be chosen with great care, and their association with one another must be carefully managed. Perhaps the most common error in defective titles, and certainly the most damaging one in terms of comprehension, is faulty syntax (word order).

What is a good title? We define it as the fewest possible words that adequately describe the contents of the paper.

Remember that the indexing and abstracting services depend heavily on the accuracy of the title, as do individual computerized literature-retrieval systems. An improperly titled paper may be virtually lost and never reach its intended audience.

Some authors mistakenly sacrifice clarity in an attempt to be witty. The title of a paper need not, and generally should not, be clever. It must, however, be clear. An example (adapted from Halm and Landon 2007): "Association between

Diuretic Use and Cardiovascular Mortality" could be an adequate title. The authors should resist the temptation to use instead "Dying to Pee."

LENGTH OF THE TITLE

Occasionally, titles are too short. A paper was submitted to the *Journal of Bacteriology* with the title "Studies on *Brucella*." Obviously, such a title was not very helpful to the potential reader. Was the study taxonomic, genetic, biochemical, or medical? We would certainly want to know at least that much.

Much more often, titles are too long. Ironically, long titles are often less meaningful than short ones. A century or so ago, when science was less specialized, titles tended to be long and nonspecific, such as "On the addition to the method of microscopic research by a new way of producing colour-contrast between an object and its background or between definite parts of the object itself" (Rheinberg J. 1896. *J. R. Microsc. Soc.* 373). That certainly sounds like a poor title; perhaps it would make a good abstract.

Not only scientists have written rambling titles. Consider this one from the year 1705: *A Wedding Ring Fit for the Finger, or the Salve of Divinity on the Sore of Humanity with directions to those men that want wives, how to choose them, and to those women that have husbands, how to use them.* Ironically, this title appeared on a miniature book (Bernard A. 1995. Now all we need is a title: famous book titles and how they got that way. New York: Norton, p. 58).

Without question, most excessively long titles contain "waste" words. Often, these waste words appear right at the start of the title, words such as "Studies on," "Investigations on," and "Observations on." An opening *A, An,* or *The* is also a waste word. Certainly, such words are useless for indexing purposes.

NEED FOR SPECIFIC TITLES

Let us analyze a sample title: "Action of Antibiotics on Bacteria." Is it a good title? In *form* it is; it is short and carries no excess baggage (waste words). Certainly, it would not be improved by changing it to "Preliminary Observations on the Effect of Certain Antibiotics on Various Species of Bacteria." However (and this brings us to the next point), most titles that are too short are too short because they include general rather than specific terms.

We can safely assume that the study introduced by the above title did *not* test the effect of *all* antibiotics on *all* kinds of bacteria. Therefore, the title is essentially meaningless. If only one or a few antibiotics were studied, they should be individually listed in the title. If only one or a few organisms were tested, they should be individually listed in the title. If the number of antibiotics or organisms was awkwardly large for listing in the title, perhaps a group name could have been substituted. Examples of more acceptable titles are the following:

"Action of Streptomycin on *Mycobacterium tuberculosis*."
"Action of Streptomycin, Neomycin, and Tetracycline on Gram-Positive Bacteria."
"Action of Polyene Antibiotics on Plant-Pathogenic Bacteria."
"Action of Various Antifungal Antibiotics on *Candida albicans* and *Aspergillus fumigatus*."

Although these titles are more acceptable than the sample, they are not especially good because they are still too general. If the "Action of" can be defined easily, the meaning might be clearer. For example, the first title might have been phrased "Inhibition of Growth of *Mycobacterium tuberculosis* by Streptomycin."

Long ago, Leeuwenhoek used the word "animalcules," a descriptive but not very specific word. In the 1930s, Howard Raistrick published an important series of papers under the title "Studies on Bacteria." A similar paper today would have a much more specific title. If the study featured an organism, the title would give the genus and species and possibly even the strain. If the study featured an enzyme in an organism, the title would not be anything like "Enzymes in Bacteria." It would be something like "Dihydrofolate Reductase Produced by *Bacillus subtilis*."

IMPORTANCE OF SYNTAX

In titles, be especially careful of syntax. Most of the grammatical errors in titles are due to faulty word order.

A paper was submitted to the *Journal of Bacteriology* with the title "Mechanism of Suppression of Nontransmissible Pneumonia in Mice Induced by Newcastle Disease Virus." Unless this author had somehow managed to demonstrate spontaneous generation, it must have been the pneumonia that was induced and not the mice. (The title should have read: "Mechanism of Suppression of Nontransmissible Pneumonia Induced in Mice by Newcastle Disease Virus.")

If you no longer believe that babies result from a visit by the stork, we offer this title (*Am. J. Clin. Pathol.* 52:42, 1969): "Multiple Infections among Newborns Resulting from Implantation with *Staphylococcus aureus* 502A." (Is this the "Staph of Life"?)

Another example (*Clin. Res.* 8:134, 1960): "Preliminary Canine and Clinical Evaluation of a New Antitumor Agent, Streptovitacin." When that dog gets through evaluating streptovitacin, we've got some work we'd like that dog to look over. A grammatical aside: Please be careful when you use "using." The word "using" might well be the most common dangling participle in scientific writing. Either there are some more smart dogs, or "using" is misused in this sentence from a manuscript: "Using a fiberoptic bronchoscope, dogs were immunized with sheep red blood cells."

Dogs aren't the only smart animals. A manuscript was submitted to the *Journal of Bacteriology* under the title "Isolation of Antigens from Monkeys Using Complement-Fixation Techniques."

Even bacteria are smart. A manuscript was submitted to the *Journal of Clinical Microbiology* under the title "Characterization of Bacteria Causing Mastitis by Gas-Liquid Chromatography." Isn't it wonderful that bacteria can use GLC?

THE TITLE AS A LABEL

The title of a paper is a label. It normally is not a sentence. Because it is not a sentence, with the usual subject-verb-object arrangement, it is simpler than a sentence (or, at least, shorter), but the order of the words becomes even more important.

Actually, a few journals do permit a title to be a sentence. An example of such a title: "Fruit Flies Diversify Their Offspring in Response to Parasite Infection" (*Science* 349:747, 2015). One might object to such a title because presence of a verb (in this case, *diversify*) makes the title seem like a loud assertion. Such a title may sound dogmatic because we are not accustomed to seeing authors present their results in the present tense, for reasons that are discussed in Chapter 30. Rosner (1990, p. 108) gave the name "assertive sentence title" (AST) to this kind of title and presented a number of reasons why such titles should not be used. In particular, ASTs are "improper and imprudent" because "in some cases the AST boldly states a conclusion that is then stated more tentatively in the summary or elsewhere" and "ASTs trivialize a scientific report by reducing it to a one-liner."

The meaning and order of the words in the title are important to the potential reader who sees the title in the journal table of contents. But these considerations are equally important to *all* potential users of the literature, including those (probably a majority) who become aware of the paper via secondary sources. Thus, the title should be useful as a label accompanying the paper itself, and it also should be in a form suitable for the machine-indexing systems used by *Chemical Abstracts*, MEDLINE, and others. In short, the terms in the title should be those that highlight the significant content of the paper.

As an aid to readers, journals commonly print *running titles* or *running heads* at the top of each page. Often the title of the journal or book is given at the top of left-facing pages and the article or chapter title is given at the top of right-facing ages (as in this book). Usually, a short version of the title is needed because of space limitations. (The maximum character count is likely to be stated in the journal's instructions to authors.) It can be wise to suggest an appropriate running title on the title page of the manuscript.

ABBREVIATIONS AND JARGON

Titles should almost never contain abbreviations, chemical formulas, proprietary (rather than generic) names, jargon, and the like. In designing the title, the author should ask: "How would I look for this kind of information in an index?" If the paper concerns an effect of hydrochloric acid, should the title include the words "hydrochloric acid," or should it contain the much shorter and readily recognizable "HCl"? The answer seems obvious. Most of us would look under "hy" in an index, not under "hc." Furthermore, if some authors used (and journal editors permitted) HCl and others used hydrochloric acid, the user of the bibliographic services might locate only part of the published literature, not noting that additional references are listed under another, abbreviated entry. Actually, the larger secondary services have computer programs that can bring together entries such as deoxyribonucleic acid, DNA, and even ADN (*acide deoxyribonucleique*). However, by far the best rule for authors (and editors) is to avoid abbreviations in titles. And the same rule should apply to proprietary names, jargon, and unusual or outdated terminology.

MORE ABOUT TITLE FORMAT

Many editors are opposed to main title-subtitle arrangements and to hanging titles. The main title-subtitle (series) arrangement was quite common some years ago. (Example: "Studies on Bacteria. IV. Cell Wall of *Staphylococcus aureus*.") Today, many editors believe that it is important, especially for the reader, that each published paper "present the results of an independent, cohesive study; thus, numbered series titles are not allowed" (instructions to authors, *Journal of Bacteriology*). Series papers, in the past, have tended to relate to each other too closely, giving only bits and pieces with each contribution; thus, the reader was severely handicapped unless the whole series could be read consecutively. Furthermore, the series system is annoying to editors because of scheduling problems and delays. (What happens when IV is accepted but III is rejected or delayed in review?) Additional objections are that a series title almost always provides considerable redundancy; the first part (before the roman numeral) is usually so general as to be useless, and the results when the secondary services spin out an index are often unintelligible. (Article titles phrased as questions also can become unintelligible, and so they probably should not be used.)

The hanging title (similar to a series title but with a colon instead of a roman numeral) is considerably better, avoiding some of the problems mentioned. Some journals, especially in the social sciences (Hartley 2007), seem to favor hanging titles, presumably on the grounds that it is helpful to get the most important words of the title up to the front. (Example: "Environmental Science in the Media: Effects of Opposing Viewpoints on Risk and Uncertainty Perceptions" *Science Communication* 37:287, 2015). Occasionally, hanging titles may aid the reader, but they may appear pedantic, emphasize the general term rather than a more significant term, necessitate punctuation, and scramble indexes.

Use of a straightforward title does not lessen the need for proper syntax, however, or for the proper form of each word in the title. For example, a title reading "New Color Standard for Biology" would seem to indicate the development of color specifications for use in describing plant and animal specimens. However, in the title "New Color Standard for Biologists" (*Bioscience* 27:762, 1977), the new standard might be useful for study of the taxonomy of biologists, permitting us to separate the green biologists from the blue ones.

CHAPTER 8

How to List the Authors and Addresses

The list of authors establishes accountability as well as credit.

—National Academies Committee on Science, Engineering, and Public Policy

THE ORDER OF THE NAMES

"If you have co-authors, problems about authorship can range from the trivial to the catastrophic" (O'Connor 1991, p. 10).

The easiest part of preparing a scientific paper is simply entering the bylines: the authors and addresses. Sometimes.

We haven't yet heard of a duel being fought over the order of listing of authors, but there have been instances in which otherwise reasonable, rational colleagues have become bitter enemies solely because they could not agree on whose names should be listed or in what order.

What is the right order? Unfortunately, there are no agreed-upon rules or generally accepted conventions. Some authors, perhaps to avoid arguments among themselves, agree to list their names alphabetically. In the field of mathematics, this practice appears to be standard. Some pairs of researchers who repeatedly collaborate take turns being listed first. If allowed by the journal, sometimes papers include a note indicating that the first two authors contributed equally to the research.

In the past, there was a general tendency to list the head of the laboratory (or, more generally, the head of the research group) as an author whether or not he or she actively participated in the research. Often, the "head" was placed last (second of two authors, third of three, etc.). As a result, the terminal spot seemed to acquire prestige. Thus, two authors, neither of whom was head of a laboratory or even necessarily a senior professor, would vie for the second spot. If there were three or more authors, the prestige-seeking author would want the first or last position, but not the one in between.

Commonly, the first author is the person who played the lead role in the research. Qualification to be listed first does not depend on rank. A graduate student, or even an undergraduate, may be listed first if he or she led the research project. And even Nobel laureates are not to be listed first unless their contributions predominate. Multiple authors may then be listed approximately in order of decreasing contribution to the work. In some fields, the head of the laboratory is still often listed last, in which case this position may continue to command particular respect. However, the head should be included only if he or she indeed at least provided guidance. In general, all those listed as authors should have been involved enough to defend the paper or a substantial aspect thereof. Some authors who did not participate substantially in the research have come to regret their inclusion when the reported research was found deficient or even fraudulent.

There is often a tendency to use the laundry-list approach, naming as an author practically everyone in the laboratory. In addition, the trend toward col-

"I DIDN'T EXACTLY WRITE THE ARTICLE, BUT... WELL, I DIDN'T EXACTLY DO THE RESEARCH, EITHER."

(© ScienceCartoonsPlus.com)

laborative research is steadily increasing. Thus, the average number of authors per paper is on the rise.

DEFINITION OF AUTHORSHIP

Perhaps we can now define authorship by saying that the listing of authors should include those, and only those, who actively contributed to the overall conceptualization, design, and execution of the research. Further, the authors should normally be listed in order of importance *to the research*. Colleagues or supervisors should neither ask to have their names on manuscripts nor allow their names to be put on manuscripts reporting research with which they themselves have not been intimately involved. An author of a paper should be defined as one who takes intellectual responsibility for the research results being reported. However, this definition must be tempered by realizing that modern science in many fields is collaborative and multidisciplinary. It may be unrealistic to assume that all authors can defend all aspects of a paper written by contributors from a variety of disciplines. Even so, each author should be held fully responsible for his or her choice of colleagues.

Admittedly, deciding on authorship is not always easy. It is often incredibly difficult to analyze the intellectual input to a paper. Certainly, those who have worked together intensively for months or years on a research problem might have difficulty in remembering who had the original research concept or whose brilliant idea was the key to the success of the experiments. And what do these colleagues do when everything suddenly falls into place as a result of a searching question by the traditional "guy in the next lab" who had nothing whatever to do with the research?

Each listed author should have made an important contribution to the study being reported, with the word *important* referring to those aspects of the study that produced new information, the concept that defines an original scientific paper.

The sequence of authors on a published paper should be decided, unanimously, before the research is started. A change may be required later, depending on which turn the research takes, but it is foolish to leave this important question of authorship to the very end of the research process.

In some fields, it is not rare to see 10 or more authors listed at the head of a paper. For example, a paper by F. Bulos and others (*Phys. Rev. Letters* 13:486, 1964) had 27 authors and only 12 paragraphs. Such papers frequently come from laboratories that are so small that 10 people couldn't fit into the lab, let alone make a meaningful contribution to the experiment. What accounts for the tendency to list a host of authors? There may be several reasons, but one of them no doubt relates to the publish-or-perish syndrome. Some workers wheedle or cajole their colleagues so effectively that they become authors of most or all of the papers coming out of their laboratory. Their research productivity might in fact be meager, yet at year's end their publication lists might indeed be extensive. In some institutions, such padded lists might result in promotion. Nonetheless, the practice is not recommended. Perhaps a few administrators are fooled, and momentary advantages are sometimes gained by these easy riders. But we suspect that *good* scientists do not allow dilution of their own work by adding other people's names for minuscule contributions, nor do good scientists want their own names sullied by addition of the names of a whole herd of lightweights.

To repeat, the scientific paper should list as authors only those who contributed *substantially* to the work. Unjustified listing of multiple authors adversely affects the *real* investigators and can lead to bibliographic nightmares. For more on issues relating to the definition of authorship, see Davidoff (2000), Claxton (2005), Scott-Lichter and the Editorial Policy Committee, Council of Science Editors (2012), and International Committee of Medical Journal Editors (2014).

DEFINING THE ORDER: AN EXAMPLE

Perhaps the following example will help clarify the level of conceptual or technical involvement that should define authorship.

Suppose that Scientist A designs a series of experiments that might result in important new knowledge, and then Scientist A tells Technician B exactly how to perform the experiments. If the experiments work out and a manuscript results, Scientist A should be the sole author, even though Technician B did all the physical work. (Of course, the assistance of Technician B should be recognized in the acknowledgments.)

Now let us suppose that the experiments just described do not work out. Technician B takes the negative results to Scientist A and says something like, "I think we might get this damned strain to grow if we change the incubation temperature from 24 to 37°C and if we add serum albumin to the medium." Scientist A agrees to a trial, the experiments this time yield the desired outcome, and a paper results. Technician B also provides some insights that contribute to the interpretation of the results. In this case, Scientist A and Technician B, in that order, should both be listed as authors.

Let us take this example one step further. Suppose that the experiments at 37°C and with serum albumin work, but that Scientist A perceives that there is now an obvious loose end; that is, growth under these conditions suggests that the test organism is a pathogen, whereas the previously published literature had indicated that this organism was nonpathogenic. Scientist A now asks colleague Scientist C, an expert in pathogenic microbiology, to test this organism for pathogenicity. Scientist C runs a quick test by injecting the test substance into laboratory mice in a standard procedure that any medical microbiologist would use and confirms pathogenicity. A few important sentences are then added to the manuscript, and the paper is published. Scientist A and Technician B are listed as authors; the assistance of Scientist C is noted in the acknowledgments.

Suppose, however, that Scientist C gets interested in this peculiar strain and proceeds to conduct a series of well-planned experiments that lead to the conclusion that this particular strain is not just mouse-pathogenic, but is the long-sought culprit in certain rare human infections. Thus, two new tables of data are added to the manuscript, and the results and discussion are rewritten. The paper is then published listing Scientist A, Technician B, and Scientist C as authors. (A case could be made for listing Scientist C as the second author.)

SPECIFYING CONTRIBUTIONS

Some journals require a list of which author or authors did what—for example, who designed the research, who gathered the data, who analyzed the data, and who wrote the paper. Some journals publish this list of contributors with the paper. Others just keep it for their own information. Sometimes, there are contributors who are not authors—for example, people who obtained some of the data but did not participate more broadly in the research or who provided technical or other guidance.

Requiring this list of contributions can have at least two advantages. First, it helps ensure that everyone listed as an author deserves to be listed—and that no one who ought to be listed has been left out. Second, if the list is published, it can help readers determine which author to contact for which type of information.

PROPER AND CONSISTENT FORM

As to names of authors, the preferred designation normally is given name, middle initial, surname. If an author uses only initials, which has been a regrettable tendency in science, the scientific literature may become confused.

If there are two people named Jonathan B. Jones, the literature services can probably keep them straight (by addresses). But if dozens of people published under the name J. B. Jones (especially if, on occasion, some of them use Jonathan B. Jones), the retrieval services have a hopeless task in keeping things neat and tidy. Many scientists resist the temptation to change their names (for example, after marriage) at least in part to avoid confusion in the literature.

Instead of given name, middle initial, and surname, wouldn't it be better to spell out the middle name? No. Again, we must realize that literature retrieval is a computerized process and that computers can be easily confused. An author with a common name (for example, Robert Jones) might be tempted to spell out his or her middle name, thinking that Robert Smith Jones is more distinctive than Robert S. Jones. However, the resulting double name is a problem. Should the computer index the author as "Jones" or "Smith Jones"? Because double names, with or without hyphens, are common, especially in England and in Latin America, this problem is not an easy one for computers (or for their programmers).

Knowing how to list one's name on an English-language scientific paper can be difficult for international authors as different languages have different formats for names, and more than one form of transliteration can exist. For authors with Chinese names, an article by Sun and Zhou (2002) offers recommendations. And for authors of a variety of national origins, style manuals can provide guidance, as can editors at journals. Whatever format a scientist chooses, he or she should use it consistently in English-language scientific papers—rather than, for example, using Shou-Chu Qian on some papers, Shouchu Qian on others, and S. Chien on still others.

In general, scientific journals do not print degrees after authors' names and do not include titles such as Dr. (You know what "B.S." means. "M.S." is More of the Same. "Ph.D." is Piled Higher and Deeper. "M.D." is Much Deeper.) However, most medical journals do list degrees after the names. Even in medical journals, however, degrees are not given in the references. Contributors should consult the journal's instructions to authors or a recent issue regarding preferred usage.

LISTING THE ADDRESSES

The principles for listing the addresses are simple but often violated. Therefore, authors cannot always be connected with addresses. Most often, however, it has been the style of the journal that creates confusion, rather than sins of commission or omission by the author.

With one author, one address is given (the name and address of the laboratory in which the work was done). If, before publication, the author has moved to a different address, the new address should be indicated in a "present address" footnote.

When two or more authors are listed, each in a different institution, the addresses should be listed in the same order as the authors.

The main problem arises when a paper is published by, let us say, three authors from two institutions. In such instances, each author's name and

THE AUTHOR LIST: GIVING CREDIT WHERE CREDIT IS DUE

The first author
Senior grad student on the project. Made the figures.

The third author
First year student who actually did the experiments, performed the analysis and wrote the whole paper. Thinks being third author is "fair".

The second-to-last author
Ambitious assistant professor or post-doc who instigated the paper.

Michaels, C., Lee, E. F., Sap, P. S., Nichols, S. T., Oliveira, L., Smith, B. S.

The second author
Grad student in the lab that has nothing to do with this project, but was included because he/she hung around the group meetings (usually for the food).

The middle authors
Author names nobody really reads. Reserved for undergrads and technical staff.

The last author
The head honcho. Hasn't even read the paper but, hey, he got the funding, and his famous name will get the paper accepted.

JORGE CHAM © 2005

www.phdcomics.com

(www.phdcomics.com)

address should include an appropriate designation such as a superior *a, b,* or *c* after the author's name and before (or after) the appropriate address. (Sometimes a journal may just request the affiliation of each author and then do the formatting itself. In this regard as in others, follow the instructions to authors.)

This convention has been useful to readers wanting to know whether an author is at Yale or at Harvard. Clear identification of authors and addresses has been important to several of the secondary services. For these services to function properly, they needed to know whether a paper published by an author with a common name was by the person with that name at Iowa State, Cornell, Cambridge University in England, or Peking University. Only when authors could be properly identified could their publications be grouped together in citation indexes.

A SOLUTION: ORCID

Even with addresses, authors can be difficult to distinguish from one another—for example, if two scientists with the same name work at the same institution. Also, some scientists move from one institution to another or do not state their names the same way on all their papers over the years, and so their work is hard to track. Fortunately, a mechanism now exists to unambiguously identify each author.

This mechanism is ORCID, which stands for "Open Researcher and Contributor ID." An ORCID identifier is a persistent identification number that you can obtain and include with your research communications. When you apply at the ORCID website, you receive a unique identification number and establish an ORCID record online. You can then associate this number with your journal articles, grant proposals, and other writings, both in the future and retroactively. Many journals now ask authors to supply their ORCID identifiers. Information about the ORCID initiative and a link through which to obtain an ORCID identifier appear at orcid.org.

PURPOSES OF THE ADDRESSES

Remember that the address serves two purposes. It helps to identify the author; it also indicates how to contact him or her. Because scientists now communicate largely by email, an email address generally should be included at least for the author to whom inquiries about the paper should be conveyed. Some journals use asterisks, footnotes, or the acknowledgments to identify this person. Authors should be aware of journal policy in this regard, and they should decide *in advance* which author will serve in this role.

The author who should receive inquiries is called the *corresponding author.* Journals ask that a corresponding author be designated for each paper. The corresponding author typically submits the paper, receives the editor's decision whether to publish it, submits revisions, works with the editorial office after acceptance (for example, by answering questions from the manuscript editor and checking page proofs), and responds to inquiries from readers. The corresponding author should be someone who expects to be readily reachable during and after the publication process. Opinions vary as to whether being a corresponding author is an honor or just a task.

Unless scientists wish to publish anonymously (or as close to it as possible), full names and a full address should be considered obligatory.

CHAPTER 9

How to Prepare the Abstract

I have the strong impression that scientific communication is being seriously hindered by poor quality abstracts written in jargon-ridden mumbo-jumbo.

—Sheila M. McNab

DEFINITION

An abstract should be viewed as a miniature version of the paper. The abstract should provide a *brief* summary of each of the main sections of the paper: introduction, materials and methods, results, and discussion. As Houghton (1975) put it, "An abstract can be defined as a summary of the information in a document."

"A well-prepared abstract enables readers to identify the basic content of a document quickly and accurately, to determine its relevance to their interests, and thus to decide whether they need to read the document in its entirety" (American National Standards Institute 1979b). The abstract should not exceed the length specified by the journal (commonly, 250 words), and it should be designed to define clearly what is dealt with in the paper. Typically, the abstract should be typed as a single paragraph, as in Figure 9.1. Some journals, however, run "structured" abstracts consisting of a few brief paragraphs, each preceded by a standardized subheading, as in Figure 9.2. Many people will read the abstract, either in the original journal or as retrieved by computer search.

The abstract should (1) state the principal objectives and scope of the investigation, (2) describe the methods employed, (3) summarize the results, and (4) state the principal conclusions. The importance of the conclusions is indicated by the fact that they are often given three times: once in the abstract, again in the introduction, and again (in more detail, probably) in the discussion.

EFFECTS OF SCIENTIFIC-WRITING TRAINING ON KNOWLEDGE AND PUBLICATION OUTPUT

(An Imaginary Study)

Scientists must write to succeed, but few receive training in scientific writing. We studied the effects of a scientific-communication lecture series, alone and combined with feedback on writing, on scientific-communication knowledge and publication performance. During the spring 2010 semester, 50 science PhD students in their last year at Northeast Southwest University were randomly assigned to receive no instruction in scientific writing, attend eight 1-hour lectures on the topic, or attend these lectures and receive feedback from classmates and an instructor on successive parts of a scientific paper they drafted. Members of each group then took a test of scientific-communication knowledge, and the publication output of each group was monitored for 5 years. Members of the groups receiving instruction scored between 80 and 98 percent on the test of scientific-communication knowledge, whereas all but two members of the control group scored below 65 percent. Although on average the group receiving lectures and feedback scored higher than the lecture-only group, the difference was not significant. During the 5-year follow-up, on average the control-group members submitted 6.1 papers to journals and had 4.1 accepted. The corresponding figures for the lecture group were 6.5 and 4.8, and those for the lecture-plus-feedback group were 8.3 and 6.7. Higher proportions of the latter two groups had papers accepted by the first journal to which they were submitted. These findings suggest that instruction in scientific writing, especially if it includes practice and feedback, can increase knowledge of scientific communication and promote publication success.

Figure 9.1. Abstract (in conventional format) of a fictional scientific paper. This abstract runs slightly less than 250 words and so would comply with typical word limits. Were a real study being reported, the statistical information probably would be more sophisticated. Note that the order of information parallels that in a typical scientific paper.

Most or all of the abstract should be written in the past tense because it refers to work done.

The abstract should never give any information or conclusion that is not stated in the paper. Literature must not be cited in the abstract (except in rare instances, such as modification of a previously published method). Likewise, normally the abstract should not include or refer to tables and figures. (Some journals, however, allow or even require the abstract to include a graphic.)

EFFECTS OF SCIENTIFIC-WRITING TRAINING ON KNOWLEDGE AND PUBLICATION OUTPUT

(An Imaginary Study)

Background. Scientists must write to succeed, but few receive training in scientific writing. We studied the effects of a scientific-communication lecture series, alone and combined with feedback on writing, on scientific-communication knowledge and publication performance.

Method. During the spring 2010 semester, 50 science PhD students in their last year at Northeast Southwest University were randomly assigned to receive no instruction in scientific writing, attend eight 1-hour lectures on the topic, or attend these lectures and receive feedback from classmates and an instructor on successive parts of a scientific paper they drafted. Members of each group then took a test of scientific-communication knowledge, and the publication output of each group was monitored for 5 years.

Results. Members of the groups receiving instruction scored between 80 and 98 percent on the test of scientific-communication knowledge, whereas all but two members of the control group scored below 65 percent. Although on average the group receiving lectures and feedback scored higher than the lecture-only group, the difference was not significant. During the 5-year follow-up, on average the control-group members submitted 6.1 papers to journals and had 4.1 accepted. The corresponding figures for the lecture group were 6.5 and 4.8, and those for the lecture-plus-feedback group were 8.3 and 6.7. Higher proportions of the latter two groups had papers accepted by the first journal to which they were submitted.

Conclusion. These findings suggest that instruction in scientific writing, especially if it includes practice and feedback, can increase knowledge of scientific communication and promote publication success.

Figure 9.2. Structured version of the abstract shown in Figure 9.1. The two abstracts are the same except for division into paragraphs and inclusion of headings. As noted, the content is fictional.

TYPES OF ABSTRACTS

The preceding rules apply to the abstracts that are used in primary journals and often without change in the secondary services (*Chemical Abstracts*, etc.). This type of abstract is often called an *informative* abstract, and it is designed to condense the paper. It can and should briefly state the problem, the method used to study the problem, and the principal data and conclusions. Often, the abstract supplants the need for reading the full paper; without such abstracts, scientists would not be able to keep up in active areas of research. (However, before citing a paper, you should read it in its entirety because some abstracts—surely not yours, though!—do not convey an entirely accurate picture of the research.) This is the type of abstract that precedes the body of the paper (thus serving as a "heading") in most journals.

Another type of abstract is the *indicative* abstract (sometimes called a descriptive abstract). This type of abstract (see Figure 9.3) is designed to indicate the subjects dealt with in a paper, much like a table of contents, making it easy for potential readers to decide whether to read the paper. However, because of the descriptive rather than substantive nature, it can seldom serve as a substitute for the full paper. Thus, indicative abstracts should not be used as "heading" abstracts in research papers, but they may be used in other types of publications, such as review papers, conference reports, and government reports. Such indicative abstracts are often of great value to reference librarians.

An effective discussion of the various uses and types of abstracts was provided by McGirr (1973, p. 4), whose conclusions are well worth repeating: "When writing the abstract, remember that it will be published by itself, and should be self-contained. That is, it should contain no bibliographic, figure, or

TEACHING OF SCIENTIFIC WRITING

(An Imaginary Review Article)

In this article we summarize and discuss the literature on teaching scientific writing. Although we focus mainly on articles in peer-reviewed journals, we also draw on material in professionally oriented magazines and newsletters and in books. First we describe methods used for the literature review, including databases searched, keywords used, and languages and dates included. Then we present information on the history of teaching scientific writing and on instructional designs reported, including single sessions, intensive short courses, and semester-long courses; examples of instruction at specific institutions and under other auspices are noted. Also addressed are the teaching of English-language scientific writing to non-native users of English, the use of distance instruction in teaching scientific writing, issues in scientific-writing instruction, and current trends in the field. Finally, we identify topics on which further research appears advisable. Supplementary materials include annotated lists of textbooks and websites useful in teaching scientific writing.

Figure 9.3. Indicative (descriptive) abstract of a fictional review article. This abstract runs about 150 words. Like a table of contents, it lists topics but does not state what is said about them.

table references. . . . The language should be familiar to the potential reader. Omit obscure abbreviations and acronyms. Write the paper before you write the abstract, if at all possible."

Unless a long term is used several times within an abstract, do not abbreviate the term. Wait and introduce the appropriate abbreviation at first use in the text (probably in the introduction).

ECONOMY OF WORDS

Occasionally, a scientist omits something important from the abstract. By far the most common fault, however, is the inclusion of extraneous detail.

A scientist once had some terribly involved theory about the relation of matter to energy. He then wrote a terribly involved paper. However, the scientist, knowing the limitations of editors, realized that the abstract of his paper would have to be short and simple if the paper was to be judged acceptable. So, he spent hours and hours honing his abstract. He eliminated word after word until, finally, all of the verbiage had been removed. What he was left with was the shortest abstract ever written: "$E=mc^2$."

Today, most scientific journals print an abstract before the main text of each paper. Because the abstract precedes the paper itself, and because the editors and reviewers like a bit of orientation, the abstract is almost always the first part of the manuscript read during the review process. Therefore, it is of fundamental importance that the abstract be written clearly and simply. If you cannot make a good impression in your abstract, your cause may be lost. Very often, the reviewer may be perilously close to a final judgment of your manuscript after reading the abstract alone. This could be because the reviewer has a short attention span (often the case). However, if by definition the abstract is simply a very short version of the whole paper, it is only logical that the reviewer will often reach a preliminary conclusion, and that conclusion is likely to be the correct one. Usually, a good abstract is followed by a good paper; a poor abstract is a harbinger of woes to come.

Because an abstract is required by most journals and because a meeting abstract is a requirement for participation in a great many national and international meetings (participation sometimes being determined on the basis of submitted abstracts), scientists should master the fundamentals of abstract preparation.

When writing the abstract, examine every word carefully. If you can tell your story in 100 words, do not use 200. Economically and scientifically, it doesn't make sense to waste words. The total communication system can afford only so much verbal abuse. Of more importance to you, the use of clear, significant words will impress the editors and reviewers (not to mention readers), whereas the use of abstruse, verbose constructions might well contribute to a check in the "reject" box on the review form.

Here's an example of an especially brief abstract, which accompanied a paper by M. V. Berry and colleagues (*J. Phys. A: Math. Theor.* 44:492001, 2011). The title of the paper: "Can apparent superluminal neutrino speeds be explained as a quantum weak measurement?" The abstract: "Probably not." Should you write abstracts this short? Well, probably not. Normally an abstract should be more informative than this one. But at least, unlike some meandering abstracts, this one answers the question that the research addressed.

AKIN TO ABSTRACTS

Some journals include, in addition to abstracts, other components briefly conveying key points to readers, skimmers, or browsers. For example, some journals ask authors to provide a bulleted list of key messages of their articles, either for posting only online or for publication as part of the article as well. Others, for instance, request a nontechnical summary or a brief statement of implications. Some journals require such items to accompany all papers submitted; others request them only for some or all of the papers accepted for publication. Be aware that you may be asked to provide, in essence, an abstract of your abstract.

CHAPTER 10

How to Write the Introduction

A bad beginning makes a bad ending.

—Euripides

GUIDELINES

Now that we have the preliminaries out of the way, we come to the paper itself. Some experienced writers prepare their title and abstract after the paper is written, even though by placement these elements come first. You should, however, have in mind (if not on paper or in the computer) a provisional title and an outline of the paper you propose to write. You should also consider the background of the audience you are writing for so that you will have a basis for determining which terms and procedures need definition or description and which do not. If you do not have a clear purpose in mind, you might go writing off in six directions at once.

It is wise to begin writing the paper while the work is still in progress. This makes the writing easier because everything is fresh in your mind. Furthermore, the writing process itself is likely to point to inconsistencies in the results or perhaps to suggest interesting sidelines that might be followed. Thus, start the writing while the experimental apparatus and materials are still available. If you have coauthors, it is wise to write up the work while they are still available to consult.

The first section of the text proper should, of course, be the introduction. The purpose of the introduction is to supply sufficient background information to allow the reader to understand and evaluate the results of the present study without needing to refer to previous publications on the topic. The introduction should also provide the rationale for the present study. Above all, you should state briefly and clearly your purpose in writing the paper. Choose references carefully to provide the most important background information. Much of the introduction should be written in present tense because you are referring primarily to your problem and the established knowledge relating to it at the start of your work.

Guidelines for a good introduction are as follows: (1) The introduction should present first, with all possible clarity, the nature and scope of the problem investigated. For example, it should indicate why the overall subject area of the research is important. (2) It should briefly review the pertinent literature to orient the reader. It also should identify the gap in the literature that the current research was intended to address. (3) It should then make clear the objective of the research. In some disciplines or journals, it is customary to state here the hypotheses or research questions that the study addressed. In others, the objective may be signaled by wording such as "in order to determine." (4) It should state the method of the investigation. If deemed necessary, the reasons for the choice of a particular method should be briefly stated. (5) Finally, in some disciplines and journals, the standard practice is to end the introduction by stating the principal results of the investigation and the principal conclusions suggested by the results.

An introduction that is structured in this way (see, for example, Figure 10.1) has a "funnel" shape, moving from broad and general to narrow and specific. Such an introduction can comfortably funnel readers into reading about the details of your research.

REASONS FOR THE GUIDELINES

The first four guidelines for a good introduction need little discussion, being reasonably well accepted by most scientist-writers, even beginning ones. It is important to keep in mind, however, that the purpose of the introduction is to introduce the paper. Thus, the first rule (definition of the problem) is the cardinal one. If the problem is not stated in a reasonable, understandable way, readers will have no interest in your solution. Even if the reader labors through your paper, which is unlikely if you haven't presented the problem in a meaningful way, he or she will be unimpressed with the brilliance of your solution. In a sense, a scientific paper is like other types of journalism. In the introduction, you should have a "hook" to gain the reader's attention. Why did you choose *that* subject, and why is it *important?*

The second, third, and fourth guidelines relate to the first. The literature review, specification of objective(s), and identification of method should be presented in such a way that the reader will understand what the problem was and how you tried to resolve it.

Although the conventions of the discipline and the journal should be followed, persuasive arguments can be made for following the fifth guideline and thus ending the abstract by stating the main results and conclusions. Do not keep the reader in suspense; let the reader follow the development of the evidence. An O. Henry surprise ending might make good literature, but it hardly fits the mold of the scientific method.

To expand on that last point: Many authors, especially beginning authors, make the mistake of holding back their more important findings until late in the paper. In extreme cases, authors have sometimes omitted important findings from the abstract, presumably in the hope of building suspense while proceeding to a well-concealed, dramatic climax. However, this is a silly gambit that, among knowledgeable scientists, goes over like a double negative at a grammarians' picnic. Basically, the problem with the surprise ending is that the readers become bored and stop reading long before they get to the punch line. "Reading a scientific article isn't the same as reading a detective story. We want to know from the start that the butler did it." (Ratnoff 1981, p. 96).

In short, the introduction provides a road map from problem to solution. This map is so important that a bit of redundancy with the abstract is often desirable.

EXCEPTIONS

Introductions to scientific papers generally should follow the guidelines that we have noted. However, exceptions exist. For example, whereas the literature review in the introduction typically should be brief and selective, journals in some disciplines favor an extensive literature review, almost resembling a review article within the paper. Some journals even make this literature review a separate section after the introduction—yielding what might be considered an ILMRAD structure.

A colleague of ours tells of reviewing an introduction drafted by a friend in another field. The introduction contained a lengthy literature review, and our colleague advised the friend to condense it. The friend followed the advice—but after she submitted the paper to a journal, the peer reviewers and editor asked her to expand the literature review. It turned out that, unknown to our colleague, her field and her friend's had different conventions in this regard. I hope that the friend kept earlier drafts (as is a good habit to follow), so she could easily restore some of what had been deleted.

In short, the conventions in your field and the requirements of your target journal take precedence. See what, if anything, the journal's instructions to authors say about the content and structure of the introduction. Also look at some papers in the journal that report research analogous to yours, and see what the introductions are like.

Introduction to an Imaginary Paper

Scientists must write to succeed, but few receive training in scientific writing. According to recent surveys, only 9 percent of scientists in the United States,[1] 5 percent of scientists in China,[2] and 3 to 12 percent of scientists attending recent international conferences[3–5] have taken a course in scientific writing. Even when briefer forms of instruction, such as workshops, are included, only about 25 percent of U.S. scientists have received formal instruction in scientific writing.[1] Discussions at a recent roundtable[6] suggest that the figure tends to be lower in other countries.

Further, relatively little information exists regarding the effectiveness of such instruction. One study[7] indicated that compared with peers without such instruction, postdoctoral fellows who had taken a scientific-writing course as graduate students felt more confident of their scientific-writing abilities and received more comments of "well written" from peer reviewers. Another study[8] suggested that the time from submission to final acceptance tended to be shorter for papers by authors who had taken a course in scientific writing. However, a third study[9] found no difference in quality of scientific papers written by early-career scientists who had completed a week-long workshop on scientific writing and those who had spent the time vacationing at a national park. The literature appears to contain little, if anything, on effects of scientific-writing instruction on knowledge or on number of publications. Likewise, it contains little or nothing on the relative effects of different forms of scientific-writing instruction.

To help address these gaps, we compared outcomes in advanced graduate students randomly assigned to receive no instruction in scientific writing, to attend a lecture series on the topic, and to attend the lecture series and receive feedback on a draft of a scientific paper. We then tested scientific-communication knowledge and monitored publication output for 5 years. Outcome measures included number of papers submitted, number of papers accepted for publication, and time from initial acceptance to publication.

Figure 10.1. Introduction to an imaginary paper on effects of scientific-writing training. This introduction, which runs about 300 words, follows the "funnel format," moving from general to specific. All content in this introduction is fictional.

CITATIONS AND ABBREVIATIONS

If you have previously published a preliminary note or abstract of the work, you should mention this (with the citation) in the introduction. If closely related papers have been or are about to be published elsewhere, you should say so in the introduction, customarily at or near the end. Such references help to keep the literature neat and tidy for those who must search it.

In addition to the preceding rules, keep in mind that your paper may well be read by people outside your narrow specialty. Therefore, in general you should define in the introduction any specialized terms or abbreviations that you will use. By doing so, you can prevent confusion such as one of us experienced in the following situation: An acquaintance who was a law judge kept referring to someone as a GC. Calling a lawyer a gonococcus (gonorrhea-causing bacterium) seemed highly unprofessional. It turned out, however, that in law, unlike in medicine, GC stands for “general counsel.”

CHAPTER 11

How to Write the Materials and Methods Section

The greatest invention of the nineteenth century was the invention of the method of invention.

—A. N. Whitehead

PURPOSE OF THE SECTION

In the first section of the paper, the introduction, you should have stated the methodology employed in the study. If necessary, you also defended the reasons for your choice of a particular method over competing methods.

Now, in "Materials and Methods" (also designated in some cases by other names, such as "Experimental Procedures"), you must give the full details. Most of this section should be written in the past tense. The main purpose of the materials and methods section is to describe (and if necessary, defend) the experimental design and then provide enough detail so that a competent worker can repeat the experiments. Other purposes include providing information that will let readers judge the appropriateness of the experimental methods (and thus the probable validity of the findings) and that will permit assessment of the extent to which the results can be generalized. Many (probably most) readers of your paper will skip this section, because they already know from the introduction the general methods you used, and they probably have no interest in the experimental detail. However, careful writing of this section is critically important because the cornerstone of the scientific method requires that your results, to be of scientific merit, must be reproducible; and, for the results to be adjudged reproducible, you must provide the basis for repetition of the experiments by others. That experiments are unlikely to be reproduced is beside the point; the potential for reproducing the same or similar results must exist, or your paper does not represent good science.

When your paper is subjected to peer review, a good reviewer will read the materials and methods section carefully. If there is serious doubt that your experiments could be repeated, the reviewer will recommend rejection of your manuscript no matter how awe-inspiring your results.

MATERIALS

For materials, include the exact technical specifications, quantities, and source or method of preparation. Sometimes it is even necessary to list pertinent chemical and physical properties of the reagents used. In general, avoid the use of trade names; use of generic or chemical names is usually preferred. This approach avoids the advertising inherent in the trade name. Besides, the nonproprietary name is likely to be known throughout the world, whereas the proprietary name may be known only in the country of origin. However, if there are known differences among proprietary products, and if these differences might be critical, then use of the trade name, plus the name of the manufacturer, is essential. When using trade names, which are usually registered trademarks, capitalize them (Teflon, for example) to distinguish them from generic names. Normally, the generic description should immediately follow the trademark; for example, one would refer to Kleenex facial tissues. In general, it is not necessary to include trademark symbols (such as ® and ™). However, some journals ask authors to do so.

Experimental animals, plants, and micro-organisms should be identified accurately, usually by genus, species, and strain designations. Sources should be listed and special characteristics (age, sex, and genetic and physiological status) described. If human subjects were used, the criteria for selection should be described, and an "informed consent" statement should be included in the manuscript. Likewise, if human or animal subjects were used, approval by the appropriate committee should be noted.

Because the value of your paper (and your reputation) can be damaged if your results are not reproducible, you must describe research materials with great care. Examine the instructions to authors of the journal to which you plan to submit the manuscript, because important specifics are often detailed there. Below is a carefully worded statement applying to cell lines and reagents. It is taken from the information for authors of *In Vitro Cellular & Developmental Biology—Animal* (known for short as *In Vitro Animal*), a journal of the Society for In Vitro Biology.

> The source of cells utilized, species, sex, strain, race, age of donor, and whether primary or established should be clearly indicated. The name, city, and state or country of the source of reagents should be stated within parentheses when first cited. Specific tests used for verification of cell lines and novel reagents should be identified. Specific tests for the presence of mycoplasmal contamination of cell lines are recommended. If these tests were not performed, this fact should be clearly stated. Other data relating to unique biological, biochemical, and/or immunological markers should also be included if available. Publication of results in *In Vitro Animal* is based on the principle that results must be verifiable. Authors are expected to make unique reagents available to qualified investigators. Authors deriving or using cell lines are encouraged to follow the UKCCCR [United Kingdom Coordinating Committee on Cancer Research] Guidelines for the Use of Cell Lines in Cancer Research in respect to validation of identity and infection-free cultures.

METHODS

For methods the usual order of presentation is chronological. Obviously, however, related methods should be described together, and straight chronological order cannot always be followed. For example, even if a particular assay was not done until late in the research, the assay method should be described along with the other assay methods, not by itself in a later part of the materials and methods section.

HEADINGS

The materials and methods section often has subheadings. To see whether subheadings would indeed be suitable—and, if so, what types are likely to be appropriate—look at analogous papers in your target journal. When possible, construct subheadings that "match" those to be used in the results section. The writing of both sections will be easier if you strive for internal consistency, and the reader will be able to grasp quickly the relationship of a particular method to the related results.

MEASUREMENTS AND ANALYSIS

Be precise. Methods are similar to cookbook recipes. If a reaction mixture was heated, give the temperature. Questions such as "how" and "how much" should be precisely answered by the author and not left for the reviewer or the reader to puzzle over.

Statistical analyses are often necessary, but your paper should emphasize the data, not the statistics. Generally, a lengthy description of statistical methods indicates that the writer has recently acquired this information and believes that the readers need similar enlightenment. Ordinary statistical methods generally should be used without comment; advanced or unusual methods may require a literature citation. In some fields, statistical methods or statistical software customarily is identified at the end of the materials and methods section.

And again, be careful of your syntax. A recent manuscript described what could be called a disappearing method. The author stated, "The radioactivity in the tRNA region was determined by the trichloroacetic acid-soluble method of Britten et al." And then there is the painful method: "After standing in boiling water for an hour, examine the flask."

NEED FOR REFERENCES

In describing the methods of the investigations, you should give (or direct readers to) sufficient details so that a competent worker could repeat the experiments. If your method is new (unpublished), you must provide *all* of the needed detail. If, however, the method has been published in a journal, the literature reference should be given. For a method well known to readers, only the literature reference is needed. For a method with which readers might not be familiar, a few words of description tend to be worth adding, especially if the journal in which the method was described might not be readily accessible.

If several alternative methods are commonly employed, it is useful to identify your method briefly as well as to cite the reference. For example, it is better to state "cells were broken by ultrasonic treatment as previously described (9)" than to state "cells were broken as previously described (9)."

TABLES AND FIGURES

When many microbial strains or mutants are used in a study, prepare strain tables identifying the source and properties of mutants, bacteriophages, plasmids, etc. The properties of multiple chemical compounds can also be presented in tabular form, often to the benefit of both the author and the reader. Tables can be used for other such types of information.

A method, strain, or the like used in only one of several experiments reported in the paper can sometimes be described in the results section. If the description is brief enough, it may be included in a table footnote or figure legend if the journal allows.

Figures also can aid in presenting methods. Examples include flow charts of experimental protocols and diagrams of experimental apparatus.

CORRECT FORM AND GRAMMAR

Do *not* make the common error of including some of the results in this section. There is only one rule for a properly written materials and methods section: Enough information must be given so that the experiments could be reproduced by a competent colleague.

A good test, by the way (and a good way to avoid rejection of your manuscript), is to give a copy of your finished manuscript to a colleague and ask if he or she can follow the methodology. It is quite possible that in reading about your materials and methods, your colleague will pick up a glaring error that you missed simply because you were too close to the work. For example, you might have described your distillation apparatus, procedure, and products with infinite care—but then neglected to define the starting material or to state the distillation temperature.

Mistakes in grammar and punctuation are not always serious; the meaning of general concepts, as expressed in the introduction and discussion, can often survive a bit of linguistic mayhem. In materials and methods, however, exact and specific items are being dealt with and precise use of English is a must. Even a missing comma can cause havoc, as in this sentence: "Employing a straight platinum wire rabbit, sheep and human blood agar plates were inoculated. . . ." That sentence was in trouble right from the start because the first word is a dangling participle. Comprehension was not totally lost, however, until the author neglected to put a comma after "wire."

Authors often are advised, quite rightly, to minimize use of passive voice. However, in the materials and methods section—as in the current paragraph—passive voice often can validly be used, for although what was done must be specified, who did it is often irrelevant. Thus, you may write, for example, "Mice were injected with . . ." rather than "I injected the mice with . . ."; "A technician injected the mice with . . ."; or "A student injected the mice with. . . ." Alternatively, you may say, for example, "We injected . . . ," even if a single member of the team did that part of the work. (Although belief persists that journals prohibit use of first person, many journals permit use of "I" and "we.")

Because the materials and methods section usually gives short, discrete bits of information, the writing sometimes becomes telescopic; details essential to the meaning may then be omitted. The most common error is to state the action without, when necessary, stating the agent of the action. In the sentence

"To determine its respiratory quotient, the organism was, . . ." the only stated agent of the action is "the organism," and we doubt that the organism was capable of making such a determination. Here is a similar sentence: "Having completed the study, the bacteria were of no further interest." Again, we doubt that the bacteria "completed the study"; if they did, their lack of "further interest" was certainly an act of ingratitude.

"Blood samples were taken from 48 informed and consenting patients . . . the subjects ranged in age from 6 months to 22 years" (*Pediatr. Res.* 6:26, 1972). There is no grammatical problem with that sentence, but the telescopic writing leaves the reader wondering just how the 6-month-old infants gave their informed consent.

And, of course, always watch for spelling errors, both in the manuscript and in the proofs. We are not astronomers, but we suspect that a word is misspelled in the following sentence: "We rely on theatrical calculations to give the lifetime of a star on the main sequence" (*Annu. Rev. Astron. Astrophys.* 1:100, 1963). Although they might have been done with a flourish, presumably the calculations were theoretical, not theatrical.

Be aware that a spell-checker can introduce such errors and therefore cannot substitute for careful proofreading. One recent example: a spell-checker's conversion of "pacemakers in dogs" to "peacemakers in dogs." We have known some dogs that could benefit from peacemakers, but we rightly suspected that this wording was not intended in writing about canine cardiology.

CHAPTER 12

How to Write the Results

Results! Why, man, I have gotten a lot of results. I know several thousand things that won't work.

—Thomas A. Edison

CONTENT OF THE RESULTS

So now we come to the core of the paper, the data. This part of the paper is called the results section.

Contrary to popular belief, you shouldn't start the results section by describing methods that you inadvertently omitted from the materials and methods section.

There are usually two ingredients of the results section. First, you should give some kind of overall description of the experiments, providing the big picture without repeating the experimental details previously provided in materials and methods. Second, you should present the data. Your results should be presented in the past tense. (See "Tense in Scientific Writing" in Chapter 30.)

Of course, it isn't quite that easy. How do you present the data? A simple transfer of data from laboratory notebook to manuscript will hardly do.

Most importantly, in the manuscript you should present representative data rather than endlessly repetitive data. The fact that you could perform the same experiment 100 times without significant divergence in results might be of considerable interest to your major professor, but editors, not to mention readers, prefer a little bit of predigestion. Aaronson (1977, p. 10) said it another way: "The compulsion to include everything, leaving nothing out, does not prove that one has unlimited information; it proves that one lacks discrimination." Exactly the same concept, and it is an important one, was stated almost a century earlier by John Wesley Powell, a geologist who served as president of the American Association for the Advancement of Science in 1888. In Powell's words: "The fool collects facts; the wise man selects them."

HOW TO HANDLE NUMBERS

If one or only a few determinations are to be presented, they should be treated descriptively in the text. Repetitive determinations should be given in tables or graphs.

Any determinations, repetitive or otherwise, should be meaningful. Suppose that, in a particular group of experiments, a number of variables were tested (one at a time, of course). Those variables that affect the reaction become determinations or data and, if extensive, are tabulated or graphed. Those variables that do not seem to affect the reaction need not be tabulated or presented; however, it is often important to define even the negative aspects of your experiments. It is often good insurance to state what you did *not* find under the conditions of your experiments. Someone else very likely may find different results under different conditions.

If statistics are used to describe the results, they should be meaningful statistics. Erwin Neter, who was editor in chief of *Infection and Immunity,* told a classic story to emphasize this point. He referred to a paper that reputedly read: "33 1/3% of the mice used in this experiment were cured by the test drug; 33 1/3% of the test population were unaffected by the drug and remained in a moribund condition; the third mouse got away."

STRIVE FOR CLARITY

The results should be short and sweet, without verbiage. Mitchell (1968) quoted Einstein as having said, "If you are out to describe the truth, leave elegance to the tailor." Although the results section is the most important part, it is often the shortest, particularly if it is preceded by a well-written materials and methods section and followed by a well-written discussion.

The results need to be clearly and simply stated because it is the results that constitute the new knowledge that you are contributing to the world. The earlier parts of the paper (introduction, materials and methods) are designed to tell why and how you got the results; the later part of the paper (discussion) is designed to tell what they mean. Obviously, therefore, the whole paper must stand or fall on the basis of the results. Thus, the results must be presented with crystal clarity.

AVOID REDUNDANCY

Do not be guilty of redundancy in the results. The most common fault is the repetition in words of what is already apparent to the reader from examining the figures and tables. Even worse is the actual presentation, in the text, of all or many of the data shown in the tables or figures. This grave sin is committed so frequently that it is commented on at length, with examples, in the chapters on how to prepare tables and illustrations (Chapters 16 and 17).

Do not be verbose in citing figures and tables. Do not say, "It is clearly shown in Table 1 that nocillin inhibited the growth of *N. gonorrhoeae*." Say, "Nocillin inhibited the growth of *N. gonorrhoeae* (Table 1)." The latter format has multiple benefits. Because it is briefer, it helps authors comply with journals' word limits. It also is more readable. It also directs attention to what is most important: the findings, not the table or figure.

Some writers go too far in avoiding verbiage, however. Such writers often fail to provide clear antecedents for pronouns, especially "it." Here is an item from a medical manuscript: "The left leg became numb at times and she walked it off. . . . On her second day, the knee was better, and on the third day it had completely disappeared." The antecedent for both "its" is presumably "the numbness," but the wording in both instances seems a result of dumbness.

A SUPPLEMENT ON SUPPLEMENTARY MATERIAL ONLINE

Increasingly, journals are electronically posting material supplementary to papers being published. Although sometimes this material regards methods, most commonly it provides information about the results. For example, additional data may be posted, or additional tables and figures may be provided online. Whether authors may submit such supplementary material, and if so how, varies among journals. Also, norms regarding what supplementary materials to provide online vary among research fields. If you think that providing supplementary material for online posting would be desirable, consult the instructions to authors of your target journal. If possible, also see what papers analogous to yours have done in this regard. Keep in mind, too, that the journal editor may ask you to place some of your material in an online supplement.

CHAPTER 13

How to Write the Discussion

It is the fault of our rhetoric that we cannot strongly state one fact without seeming to belie some other.

—Ralph Waldo Emerson

DISCUSSION AND VERBIAGE

The discussion (which some journals term a comment, especially for short papers) is harder to define than the other sections. Thus, it is usually the hardest section to write. And, whether you know it or not, *many* papers are rejected by journal editors because of a faulty discussion, even though the data of the paper might be both valid and interesting. Even more likely, the true meaning of the data may be completely obscured by the interpretation presented in the discussion, again resulting in rejection.

Many, if not most, discussion sections are too long and verbose. As Doug Savile said, "Occasionally, I recognize what I call the squid technique: the author is doubtful about his facts or his reasoning and retreats behind a protective cloud of ink" (*Tableau,* September 1972). Another reason some discussions are long and hard to follow is that many authors think they must avoid first person. If you mean "I found that . . ." or "We conclude that, . . ." say so. Try to avoid wordier, and sometimes more ambiguous, constructions such as "It was found in the present investigation that . . ." and "It is concluded that."

Some discussion sections remind one of the diplomat, described by Allen Drury in *Advise and Consent* (Garden City, NY: Doubleday, 1959, p. 47), who characteristically gave "answers which go winding and winding off through the interstices of the English language until they finally go shimmering away altogether and there is nothing left but utter confusion and a polite smile."

COMPONENTS OF THE DISCUSSION

What are the essential features of a good discussion? The main components will be provided if the following injunctions are heeded.

1. Try to present the principles, relationships, and generalizations shown by the results. And bear in mind, in a good discussion, *you discuss—you do not recapitulate*—the results.
2. Point out any exceptions or any lack of correlation and define unsettled points. Never take the high-risk alternative of trying to cover up or fudge data that do not quite fit.
3. Show how your results and interpretations agree (or contrast) with previously published work.
4. Don't be shy; discuss the theoretical implications of your work, as well as any possible practical applications.
5. State your conclusions as clearly as possible.
6. Summarize your evidence for each conclusion. Or, as the wise old scientist will tell you, "Never assume anything except a 4-percent mortgage."

Much as the methods and the results should correspond to each other, the introduction and the discussion should function as a pair. At least implicitly, the introduction should have posed one or more questions. The discussion should indicate what the findings say about the answers. Failure to address the initial questions commonly afflicts discussions. Be sure the discussion answers what the introduction asked.

Whereas the content of the introduction commonly moves from the general topic to your specific research, in sort of a funnel format, the discussion tends to do largely the reverse, much like an inverted funnel. For example, a well-structured discussion may first restate the main findings, then discuss how they relate to findings of previous research, then note implications and applications, and perhaps then identify unanswered questions well suited for future research. In the introduction, you invited readers into your research venue; in the discussion, you usher them out, now well informed about your research and its meaning.

FACTUAL RELATIONSHIPS

In simple terms, the primary purpose of the discussion is to show the relationships among observed facts. To emphasize this point, the story may be told about the biologist who trained a flea.

After training the flea for many months, the biologist was able to get a response to certain commands. The most gratifying of the experiments was the one in which the professor would shout the command "Jump," and the flea would leap into the air each time the command was given.

The professor was about to submit this remarkable feat to posterity via a scientific journal, but he—in the manner of the true scientist—decided to take his experiments one step further. He sought to determine the location of the receptor organ involved. In one experiment, he removed the legs of the flea, one at a time. The flea obligingly continued to jump upon command, but as each successive leg was removed, its jumps became less spectacular. Finally, with the removal of its last leg, the flea remained motionless. Time after time the command failed to get the usual response.

The professor decided that at last he could publish his findings. He set pen to paper and described in meticulous detail the experiments executed over the preceding months. His conclusion was one intended to startle the scientific world: *When the legs of a flea are removed, the flea can no longer hear.*

Claude Bishop, the dean of Canadian science editors, told a similar story. A science teacher set up a simple experiment to show her class the danger of alcohol. She set up two glasses, one containing water, the other containing gin. Into each she dropped a worm. The worm in the water swam merrily around. The worm in the gin quickly died. "What does this experiment prove?" she asked. A student from the back row piped up: "It proves that if you drink gin you won't have worms."

NOTING STRENGTHS AND LIMITATIONS

The discussion is a place to note substantial strengths and limitations of research being reported. Some authors feel awkward about including such content. However, doing so can aid readers, and it can help show editors and referees (peer reviewers) that your work is publishable.

Some authors consider it immodest to note strengths of their work—for example, superior experimental techniques, large sample size, or long follow-up. However, such information can aid readers in determining how definitive the findings are. It also can help persuade peer reviewers and editors that your work deserves publication.

What if research had significant limitations—such as difficulties with a technique, a relatively small sample size, or relatively short follow-up? Some authors might try to hide such limitations. However, doing so runs counter to the openness that should characterize science. And astute reviewers, editors, or readers might well notice the limitations—and assume, either to themselves or in writing, that you were too naïve to notice them. It is better, therefore, to identify substantial limitations yourself. In doing so, you may be able to discuss what impact, if any, the limitations are likely to have on the conclusions that can be drawn.

Not every discussion needs to discuss strengths or limitations of the research. However, if research has strengths or limitations major enough to be worthy of note, consider addressing them in the discussion.

SIGNIFICANCE OF THE PAPER

Too often, the *significance* of the results is not discussed or not discussed adequately. If the reader of the paper finds himself or herself asking "So what?" after reading the discussion, the chances are that the author became so engrossed with the trees (the data) that he or she didn't really notice how much sunshine had appeared in the forest.

The discussion should end with a short summary or conclusion regarding the significance of the work. (In some journals, papers include a separate conclusion section.) We like the way Anderson and Thistle (1947) said it: "Finally, good writing, like good music, has a fitting climax. Many a paper loses much of its effect because the clear stream of the discussion ends in a swampy delta." Or, in the words of T.S. Eliot, many scientific papers end "Not with a bang but a whimper."

DEFINING SCIENTIFIC TRUTH

In showing the relationships among observed facts, you do not need to reach cosmic conclusions. Seldom will you be able to illuminate the whole truth; more often, the best you can do is shine a spotlight on one area of the truth. Your one area of truth can be illuminated by your data; if you extrapolate to a bigger picture than that shown by your data, you may appear foolish to the point that even your data-supported conclusions are cast into doubt.

One of the more meaningful thoughts in poetry was expressed by Sir Richard Burton in *The Kasidah:*

> All Faith is false, all Faith is true;
> Truth is the shattered mirror strown
> In myriad bits; while each believes
> His little bit the whole to own.

So exhibit your little piece of the mirror, or shine a spotlight on one area of the truth. The "whole truth" is a subject best left to the ignoramuses, who loudly proclaim its discovery every day.

When you describe the meaning of your little bit of truth, do it simply. The simplest statements evoke the most wisdom; verbose language and fancy technical words are used to convey shallow thought.

CHAPTER 14

How to State the Acknowledgments

Life is not so short but that there is always time enough for courtesy.

—Ralph Waldo Emerson

INGREDIENTS OF THE ACKNOWLEDGMENTS

The main text of a scientific paper is usually followed by two additional sections, namely, the acknowledgments and the references.

As to the acknowledgments, two possible ingredients require consideration.

First, you should acknowledge any significant technical help that you received from any individual, whether in your laboratory or elsewhere. You should also acknowledge the source of special equipment, cultures, or other materials. You might, for example, say something like, "Thanks are due to J. Jones for assistance with the experiments and to R. Smith for valuable discussion." (Of course, most of us who have been around for a while recognize that this is simply a thinly veiled way of admitting that Jones did the work and Smith explained what it meant.)

Second, it is usually the acknowledgments wherein you should acknowledge any outside financial assistance, such as grants, contracts, or fellowships. (In this time of scarce funding, we can be especially appreciative of such support.)

BEING COURTEOUS

The important element in acknowledgments is simple courtesy. There isn't anything really scientific about this section of a scientific paper. The same rules that would apply in any other area of civilized life should apply here. If you borrowed a neighbor's lawn mower, you would (we hope) remember to say thanks for it. If your neighbor gave you a really good idea for landscaping your property and you then put that idea into effect, you would (we hope) remember to say thank you. It is the same in science; if your neighbor (your colleague) provided important ideas, important supplies, or important equipment, you should thank him or her. And you must say thanks *in print,* because that is the way that scientific landscaping is presented to its public.

A word of caution is in order. Before mentioning someone in an acknowledgment, you should obtain permission from him or her. Often, it is wise to show the proposed wording of the acknowledgment to the person whose help you are acknowledging. He or she might well believe that your acknowledgment is insufficient or (worse) that it is too effusive. If you have been working so closely with an individual that you have borrowed either equipment or ideas, that person is most likely a friend or valued colleague. It would be silly to risk either your friendship or the opportunities for future collaboration by placing in print a thoughtless word that might be offensive. An inappropriate thank-you can be worse than none at all, and if you value the advice and help of friends and colleagues, you should be careful to thank them in a way that pleases rather than displeases them.

Furthermore, if your acknowledgment relates to an idea, suggestion, or interpretation, be very specific about it. If your colleague's input is too broadly stated, he or she could well be placed in the sensitive and embarrassing position of having to defend the entire paper. Certainly, if your colleague is not a coauthor, you must not make him or her a responsible party to the basic considerations treated in your paper. Indeed, your colleague may not agree with some of your central points, and it is not good science and not good ethics for you to phrase the acknowledgments in a way that seemingly denotes endorsement.

We wish that the word "wish" would disappear from acknowledgments. Wish is a perfectly good word when you mean wish, as in "I wish you success." However, if you say "I wish to thank John Jones," you are wasting words. You may also be introducing the implication that "I wish that I could thank John Jones for his help but it wasn't all that great." "I thank John Jones" is sufficient.

CHAPTER 15

How to Cite the References

Manuscripts containing innumerable references are more likely a sign of insecurity than a mark of scholarship.

—William C. Roberts

RULES TO FOLLOW

There are two rules to follow in the references section, just as in the acknowledgments section.

First, list only significant published references. References to unpublished data, abstracts, theses, and other secondary materials should not clutter up the references or literature-cited section. If such a reference seems essential, you may add it parenthetically or, in some journals, as a footnote in the text. A paper that has been accepted for publication can be listed in the literature cited, citing the name of the journal followed by "in press" or "forthcoming."

Second, ensure that all parts of every reference are accurate. Doing so may entail checking every reference against the original publication before the manuscript is submitted and perhaps again at the proof stage. Take it from an erstwhile librarian: There are far more mistakes in the references section of a paper than anywhere else.

Don't forget, as a final check, to ensure that all references cited in the text are indeed listed in the literature cited and that all references listed in the literature cited are indeed cited somewhere in the text.

ELECTRONIC AIDS TO CITATION

Checking that every reference is accurate, and that all cited items appear in the reference list, has become much easier in the electronic era. Common word-processing programs include features for tasks such as creating, numbering, and formatting footnotes and endnotes. These features can aid in citing references and developing reference lists. Some journals, however, say not to use these features, which can interfere with their publishing process. Check the journal's instructions to authors in this regard.

Perhaps more notably, citation-management software—such as EndNote, Reference Manager, and RefWorks—lets a researcher develop a database of references and use it to create reference lists in the formats of many journals. Rather than keying in the information on each reference, you may be able to import it from bibliographic databases. Once the information is accurately entered, it should remain correct whenever it appears in a reference. Do, however, still check references. Electronic gremlins sometimes lurk. So does human error; if somehow you indicated the wrong reference, the wrong reference will appear.

If you are not using reference-management software, consider looking into doing so. Using such software can especially save you time if you will cite some of the same references in multiple publications or if journals in your field have a variety of reference styles. If you study or work at a university or other research institution, you might easily be able to obtain such software through it. Also, some universities provide instruction in using such software, for example, through their libraries. Consider checking.

CITATIONS IN THE TEXT

Many authors use slipshod methods in citing literature. A common offender is the "handwaving reference," in which the reader is glibly referred to "Smith's elegant contribution" without any hint of what Smith reported or how Smith's results relate to the present author's results. If a reference is worth citing, the reader should be told why.

Even worse is the nasty habit some authors have of insulting the authors of previous studies. It is probably all right to say "Smith (2015) did not study. . . ." But it is not all right to say "Smith (2015) totally overlooked, . . ." "Smith (2015) ignored, . . ." or "Smith (2015) failed to. . . ."

Some authors get into the habit of putting all citations at the end of sentences. This is wrong. The reference should be placed at that point in the sentence to which it applies. Michaelson (1990) gave this example:

> We have examined a digital method of spread-spectrum modulation for multiple-access satellite communication and for digital mobile radiotelephony.[1,2]

Note how much clearer the citations become when the sentence is recast as follows:

> We have examined a digital method of spread-spectrum modulation for use with Smith's development of multiple-access communication[1] and with Brown's technique of digital mobile radiotelephony.[2]

REFERENCE STYLES

Journals vary considerably in their style of handling references. O'Connor (1978) looked at 52 scientific journals and found 33 different styles for listing references. Some journals include article titles within references, and some do not. Some insist on inclusive pagination, whereas others print the numbers of first pages only.

If you use an electronic reference management system, and if that system includes the styles of all the journals in which you might like to publish, you might not need to concern yourself in detail with differences among reference styles. In that case, perhaps just skim—or even skip—the sections of this chapter that discuss formats for citing and listing references. If, however, you might at least occasionally be preparing and citing references by traditional means, we advise you to read these sections.

Whether electronically or otherwise, the smart author retains full information about every item that might be cited. Then, in preparing a manuscript, he or she has all the needed information. It is easy to edit out information; it is indeed laborious to track down 20 or so references to add article titles or ending pages when a journal editor requires you to do so. Even if you know that the journal to which you plan to submit your manuscript uses a short form (no article titles, for example), you would still be wise to establish your reference list in the complete form. This is good practice because (1) the journal you selected may reject your manuscript, and you may then decide to submit the manuscript to another journal, perhaps one with more demanding requirements, and (2) it is more than likely that you will use some of the same references again, in later research papers, review papers (and most review journals demand *full* references), or books. When you submit a manuscript for publication, make sure that the references are presented according to the instructions for authors. If the references are radically different, the editor and referees may assume that this is a sign of previous rejection or, at best, obvious evidence of lack of care.

Although there is an almost infinite variety of reference styles, most journals cite references in one of three general ways that may be referred to as name and year, alphabet-number, and citation order.

Name and Year System

The name and year system (often called the Harvard system) has been very popular for many years and is used by many journals and books, including this one. Disciplines in which it is popular include the social sciences. Its big advantage is convenience to the author. Because the references are unnumbered, references can be added or deleted easily. No matter how many times the reference list is modified, "Smith and Jones (2015)" remains exactly that. If there are two or more "Smith and Jones (2015)" references, the problem is easily handled by listing the first as "Smith and Jones (2015*a*)," the second as "Smith and Jones (2015*b*)," and so on. The disadvantages of name and year relate to readers and publishers. The disadvantage to the reader occurs when (often in the introduction) many references must be cited within one sentence or paragraph. Sometimes the reader must jump over several lines of parenthetical references before he or she can again pick up the text. Even two or three references, cited together, can distract the reader. The disadvantage to the publisher is obvious: increased cost. When "Higginbotham, Hernandez, and Chowdhary (2015)" can be converted to "(7)," printing costs can be reduced.

Because some papers are written by an unwieldy number of authors, most journals that use name and year have an "et al." (meaning "and others") rule. Commonly, it works as follows. Names are always used in citing papers with either one or two authors; for example, "Smith (2015)," "Smith and Jones (2015)." If the paper has three authors, list all three the first time the paper is cited, for example, "Smith, Jones, and Nguyen (2015)." If the same paper is cited again, it can be shorted to "Smith et al. (2015)." When a cited paper has four or more authors, it should be cited as "Smith et al. (2015)" even in the first citation. In the references section, some journals prefer that all authors be listed (no matter how many); other journals cite only the first three authors and follow with "et al."

Alphabet-Number System

This system, citation by number from an alphabetized list of references, is a modification of the name and year system. Citation by numbers keeps printing expenses within bounds; the alphabetized list, particularly if it is long, is relatively easy for authors to prepare and readers (especially librarians) to use.

Some authors who have habitually used name and year tend to dislike the alphabet-number system, claiming the citation of numbers cheats the reader. The reader should be told, the argument goes, the name of the person associated with the phenomenon; sometimes, the reader should also be told the date, on the grounds that a 1914 reference might be viewed differently than a 2014 reference.

Fortunately, these arguments can be overcome. As you cite references in the text, decide whether names or dates are important. If they are not (as is usually the case), use only the reference number: "Pretyrosine is quantitatively converted to phenylalanine under these conditions (13)." If you want to feature the name of the author, do it within the context of the sentence: "The role of the carotid sinus in the regulation of respiration was discovered by Heymans (13)." If you want to feature the date, you can also do that within the sentence: "Streptomycin was first used in the treatment of tuberculosis in 1945 (13)."

Citation Order System

The citation order system is simply a system of citing the references (by number) in the order in which they appear in the paper. This system avoids the substantial printing expense of the name and year system, and readers often like it because they can quickly refer to the references, if they so desire, in one-two-three order as they come to them in the text. It is a useful system for a journal that is basically a "note" journal, each paper containing only a few references. For long papers, with many references, citation order might not be a good system. It might not be good for the author because of the substantial renumbering chore that can result from adding or deleting references. It might not be ideal for the reader, because the non-alphabetical presentation of the reference list may result in separation of various references to works by the same author.

The first edition of this book (Day 1979, p. 40) stated that the alphabet-number system "seems to be slowly gaining ascendancy." Soon thereafter, however, the first version of the "Uniform Requirements for Manuscripts Submitted to Biomedical Journals" appeared, advocating the citation order system for the cooperating journals. Several hundred biomedical journals have adopted the "Uniform Requirements," which have evolved over the years and have now been retitled "Recommendations for the Conduct, Reporting, Editing, and Publication of Scholarly Work in Medical Journals" (International Committee of Medical Journal Editors 2014). Thus, it is not now clear which citation system, if any, will gain ascendancy. The "Uniform Requirements" document, as it still is often known, is impressive in so many ways that it has long had a powerful impact. It is in substantial agreement with a standard prepared by the American National Standards Institute (1977). With regard to literature citation, however, other usage also remains strong.

TITLES AND INCLUSIVE PAGES

Should article titles be given in references? Normally, you must follow the style of the journal; if the journal allows a choice (and some do), we recommend that you give *complete* references. By denoting the overall subjects, the article titles make it simple for interested readers (and librarians) to decide whether they need to consult none, some, or all of the cited references.

The use of inclusive pagination (first and last page numbers) makes it easy for potential users to distinguish between one-page notes and 50-page review articles. Users may wish to proceed differently depending on the number of pages involved.

JOURNAL ABBREVIATIONS

Although journal styles vary widely, one aspect of reference citation has been standardized: abbreviations of journal names. As the result of widespread adoption of a standard (American National Standards Institute 1969), almost all of the major journals and secondary services now use the same system of abbreviation. Previously, most journals abbreviated journal names (significant printing expense can be avoided by abbreviation), but there was no uniformity. The *Journal of the American Chemical Society* was variously abbreviated to "J. Amer. Chem. Soc.," "Jour. Am. Chem. Soc.," "J.A.C.S.," and so forth. These differing systems posed problems for authors and publishers alike. Now there is essentially only one system, and it is uniform. The word "Journal" is now always abbreviated "J." (Some journals omit the periods after the abbreviations.) By noting a few of the rules, authors can abbreviate many journal titles, even unfamiliar ones, without referring to a source list. It is helpful to know, for example, that all "ology" words are abbreviated at the "l." ("Bacteriology" is abbreviated "Bacteriol."; "Physiology" is abbreviated "Physiol.," etc.) Thus, if one memorizes the abbreviations of words commonly used in titles, most journal titles can be abbreviated with ease. An exception to remember is that one-word titles (*Science, Biochemistry*) are never abbreviated.

Appendix 1 lists the abbreviations for commonly used words in periodical titles. If you are unsure how to abbreviate a journal title, you can often discern the correct abbreviation from a listing in a bibliographic database, from information in the journal, or from a previous citation of the journal. Abbreviations for the titles of many journals in the biomedical sciences and related fields can be obtained from the PubMed journals database (www.ncbi.nlm.nih.gov/nlm catalog/journals).

SOME TRENDS IN REFERENCE FORMAT

Not all journals abbreviate journal titles in references. For example, APA style (*Publication Manual of the American Psychological Association* 2010, p. 185) calls for stating periodical titles in full. More generally, journals may increasingly be including full journal titles in references. Earlier, when journals appeared only in print, publishers favored abbreviating journal titles because it saved valuable space, thus saving paper costs or allowing more papers to be published. Today, with many journals appearing mainly or solely online, the space saved may be less of a consideration than are convenience to authors and clarity to readers. Writing out journal titles in full may serve especially well in journals that publish interdisciplinary papers and thus have readers who might not understand the abbreviations of some words in the titles of cited journals.

If a journal article has been published online, either exclusively or as well as in print, the publisher may have assigned it a Digital Object Identifier (DOI), which specifies a persistent link to its location on the Internet. If an article has a DOI, commonly it appears on the first page. Some reference formats include providing the DOI, if one exists, at the end of the reference. Additional information about DOIs is available at www.doi.org.

EXAMPLES OF DIFFERENT REFERENCE STYLES

So that you can see at a glance the differences among the three main systems of referencing, here are three sample references as they might appear in the references section of a journal. (In some journals, references in these systems will look somewhat different from those below because journals differ among themselves in items such as how, if at all, they use italics and boldface in references.)

Name and Year System

Álvarez GA, Suter D, and Kaiser R. 2015. Localization-delocalization transition in the dynamics of dipolar-coupled nuclear spins. Science 349:846–848.

Bern C. 2015. Chagas' disease. N. Engl. J. Med. 373:456–466.

Shipman WM. 2015. Handbook for science public information officers. Chicago: University of Chicago Press.

Alphabet-Number System

1. Álvarez, G. A., D. Suter, and R. Kaiser. 2015. Localization-delocalization transition in the dynamics of dipolar-coupled nuclear spins. Science 349:846–848.
2. Bern, C. 2015. Chagas' disease. N. Engl. J. Med. 373:456–466.
3. Shipman, W. M. 2015. Handbook for science public information officers. Chicago: University of Chicago Press.

Citation Order System

1. Bern C. Chagas' disease. N Engl J Med. 2015;373:456–66.
2. Shipman WM. Handbook for science public information officers. Chicago: University of Chicago Press, 2015.
3. Álvarez GA, Suter D, Kaiser R. Localization-delocalization transition in the dynamics of dipolar-coupled nuclear spins. Science. 2015;349:846–48.

In addition to its non-alphabetical arrangement of references, the citation order system differs from the others in its advocacy of eliminating periods after abbreviations (of journal titles, for example), periods after authors' initials, and commas after authors' surnames.

CITING ELECTRONIC MATERIAL

The Internet increasingly contains material appropriate for citation. In particular, many scientific papers now are appearing in electronic journals or being posted online as well as appearing in print. In addition, some reports, databases, and other items accessed online can be valid to cite.

Accordingly, formats have been developed, and are continuing to be developed, for citing electronic materials. These formats appear in recent editions of style manuals and in the instructions to authors of some journals. If you wish to cite electronic material, begin by consulting the instructions to authors of your target journal. These instructions may show the format(s) to use or direct you to a source of guidance in print or online. Also, you may find it useful to look in the journal for examples of references listing electronic materials.

ONE MORE REASON TO CITE CAREFULLY

Accurate citation is part of being a rigorous researcher. Whether you use reference management software or prepare references by traditional means, ensure that the right reference is cited in the right place, that all information in every citation is accurate, and that content from the cited sources is accurately reported. Such accuracy is important in ensuring that your paper is useful to readers.

And, on a very practical note, careful citation helps keep you from alienating those evaluating your paper. Commonly, some of the referees (peer reviewers) chosen by editors are researchers whose work your paper cites. If your reference section lists their writings inaccurately, or if your text misrepresents their findings or conclusions, they might well question whether you are a careful researcher.

So, take the same care with your references that you do with other aspects of your work. The effort is likely to serve you well.

PART III

Preparing the Tables and Figures

CHAPTER 16

How to Design Effective Tables

A tabular presentation of data is often the heart or, better, the brain, of a scientific paper.

—Peter Morgan

WHEN TO USE TABLES

Before proceeding to the "how to" of tables, let us first examine the question of "whether to."

As a rule, do not construct a table unless repetitive data *must* be presented. There are two reasons for this general rule. First, it is simply not good science to regurgitate reams of data just because you have them in your laboratory notebooks; only samples and breakpoints need be given. Second, the cost of publishing tables can be high compared with that of text, and all of us involved with the generation and publication of scientific literature should worry about the cost.

If you made (or need to present) only a few determinations, give the data in the text. Tables 16.1 and 16.2 are useless, yet they are typical of many tables that are submitted to journals.

Table 16.1 is faulty because two of the columns give standard conditions, not variables and not data. If temperature is a variable in the experiments, it can have its column. If all experiments were done at the same temperature, however, this single bit of information should be noted in the materials and methods section and perhaps as a footnote to the table, but not in a column in the table. The data presented in the table can be given in the text itself in a form that is readily comprehensible to the reader, without taking up space with a table. Very simply, these results would read: "Aeration of the growth medium

Table 16.1. Effect of aeration on growth of *Streptomyces coelicolor*

Temp (°C)	No. of expt	Aeration of growth medium	Growth[a]
24	5	+[b]	78
24	5	–	0

[a]As determined by optical density (Klett unit).
[b]Symbols: +, 500-ml Erlenmeyer flasks were aerated by having a graduate students blow into the bottles for 15 min out of each hour; –, identical test conditions, except that the aeration was provided by an elderly professor.

Table 16.2. Effect of temperature on growth of oak (*Quercus*) seedlings[a]

Temp (°C)	Growth in 48 h (mm)
−50	0
−40	0
−30	0
−20	0
−10	0
0	0
10	0
20	7
30	8
40	1
50	0
60	0
70	0
80	0
90	0
100	0

[a]Each individual seedling was maintained in an individual round pot, 10 cm in diameter and 100 cm high, in a rich growth medium containing 50% Michigan peat and 50% dried horse manure. Actually, it wasn't "50% Michigan"; the peat was 100% "Michigan," all of it coming from that state. And the manure wasn't half-dried (50%), it was all dried. And, come to think about it, I should have said "50% dried manure (horse)"; I didn't dry the horse at all.

was essential for the growth of *Streptomyces coelicolor.* At room temperature (24°C), no growth was evident in stationary (unaerated) cultures, whereas substantial growth (OD, 78 Klett units) occurred in shaken cultures."

Table 16.2 has no columns of identical readings, and it looks like a good table. But is it? The independent variable column (temperature) looks reasonable enough, but the dependent variable column (growth) has a suspicious number of zeros. You should question any table with a large number of zeros (whatever the unit of measurement) or a large number of 100s when percentages are used. Table 16.2 is a useless table because all it tells us is that "The oak seedlings grew at temperatures between 20 and 40°C; no measurable growth occurred at temperatures below 20°C or above 40°C."

In addition to zeros and 100s, be suspicious of plus and minus signs. Table 16.3 is of a type that often appears in print, although it is obviously not very informative. All this table tells us is that "*S. griseus, S. coelicolor, S. everycolor,* and *S. rainbowensky* grew under aerobic conditions, whereas *S. nocolor* and *S. greenicus* required anaerobic conditions." Whenever a table, or columns within a table, can be readily put into words, do it.

Some authors believe that all numerical data must be put in a table. Table 16.4 is a sad example. It gets sadder when we learn (at the end of the footnote) that the results were not statistically significant anyway ($P = 0.21$). If these data

Table 16.3. Oxygen requirement of various species of *Streptomyces*

Organism	Growth under aerobic conditions[a]	Growth under unaerobic conditions
Streptomyces griseus	+	–
S. coelicolor	+	–
S. nocolor	–	+
S. everycolor	+	–
S. greenicus	–	+
S. rainbowensky	+	–

[a]See Table 16.1 for explanation of symbols. In this experiment, the cultures were aerated by a shaking machine (New Brunswick Shaking Co., Scientific, NJ).

Table 16.4. Bacteriological failure rates

Nocillin	K Penicillin
5/35 (14)[a]	9/34 (26)

[a]Results expressed as number of failures/total, which is then converted to a percentage (within parentheses). $P = 0.21$.

Table 16.5. Adverse effects of nicklecillin in 24 adult patients

No. of patients	Side effect
14	Diarrhea
5	Eosinophilia (≥5 eos/mm^3)
2	Metallic taste[a]
1	Yeast vaginitis[b]
1	Mild rise in urea nitrogen
1	Hematuria (8–10 rbc/hpf)

[a]Both of the patients who tasted metallic worked in a zinc mine.
[b]The infecting organism was a rare strain of *Candida albicans* that causes vaginitis in yeasts but not in humans.

were worth publishing (which seems doubtful), one sentence in the results would have done the job: "The difference between the failure rates—14 percent (5 of 35) for nocillin and 26 percent (9 of 34) for potassium penicillin V—was not significant ($P=0.21$)."

In presenting numbers, give only significant figures. Nonsignificant figures may mislead the reader by creating a false sense of precision; they also make comparison of the data more difficult. Unessential data, such as laboratory numbers, results of simple calculations, and columns that show no significant variations, should be omitted.

Another very common but often useless table is the word list. Such a list might be suitable for a slide in a presentation, but it does not belong in a scientific paper. Table 16.5 is an example. This information could easily be presented in the text. A good copy editor will kill this kind of table and incorporate the data into the text. Yet, when copy editors do so (and this leads to the next rule about tables), they often find that much or all of the information was already in the text. Thus, the rule: Present the data in the text, or in a table, or in a figure. *Never* present the same data in more than one way. Of course, selected data can be singled out for discussion in the text.

Tables 16.1 to 16.5 provide typical examples of the kinds of material that should not be tabulated. Now let us look at material that should be tabulated.

HOW TO ARRANGE TABULAR MATERIAL

Having decided to tabulate, you ask yourself the question: "How do I arrange the data?" Since a table has both left-right and up-down dimensions, you have two choices. The data can be presented either horizontally or vertically. But *can* does not mean *should;* the data should be organized so that the like elements read *down,* not across.

Examine Tables 16.6 and 16.7. They are equivalent, except that Table 16.6 reads across, whereas Table 16.7 reads down. To use an old fishing expression, Table 16.6 is "bass ackward." Table 16.7 is the preferred format because it allows the reader to grasp the information more easily, and it is more compact and thus less expensive to print. The point about ease for the reader would seem to be obvious. (Did you ever try to add numbers that were listed horizontally rather than vertically?) The point about reduced printing costs refers to the fact that all columns must be wide or deep in the across arrangement because of the diversity of elements, whereas some columns (especially those with numbers) can be narrow without runovers in the down arrangement. Thus, Table 16.7 appears to be smaller than Table 16.6, although it contains the same information.

Table 16.6. Characteristics of antibiotic-producing *Streptomyces*

Determination	*S. fluoricolor*	*S. griseus*	*S. coelicolor*	*S. nocolor*
Optimal growth temp (°C)	–10	24	28	92
Color of mycelium	Tan	Gray	Red	Purple
Antibiotic produced	Fluoricillinmycin	Streptomycin	Rholmondelay[a]	Nomycin
Yield of antibiotic (mg/ml)	4,108	78	2	0

[a]Pronounced "Rumley" by the British.

Table 16.7. Characteristics of antibiotic-producing *Streptomyces*

Organism	Optimal growth temp (°C)	Color of mycelium	Antibiotic produced	Yield of antibiotic (mg/ml)
S. fluoricolor	–10	Tan	Fluoricillinmycin	4,108
S. griseus	24	Gray	Streptomycin	78
S. coelicolor	28	Red	Rholmondelay[a]	2
S. nocolor	92	Purple	Nomycin	0

[a]Where the flying fishes play.

Words in a column are lined up on the left. Numbers are lined up on the right (or on the decimal point). Table 16.7, for example, illustrates this point.

Table 16.8 is an example of a well-constructed table. It reads down, not across. Its title and headings are clear enough for readers to understand the data without referring to the text. Items in the body of the table appear in a logical

Table 16.8. Hospitalizations and total charges for neglected tropical diseases and malaria, United States, 2003–2012*

	Hospitalizations		Total charges	
Disease	No. (SE)	95% CI	US$, millions (SE)	95% CI
Cysticercosis	23,266 (778)	21,741–24,792	1,149 (56)	1,039–1,259
Malaria	14,319 (434)	13,469–15,169	387 (18)	351–423
Echinococcosis	3,919 (170)	3,586–4,252	206 (16)	174–237
Soil-transmitted helminth–associated infections	3,256 (151)	2,959–3,552	201 (19)	162–239
Dengue	2,644 (135)	2,379–2,909	89 (9)	70–107
Leprosy	2,055 (135)	1,791–2,319	94 (9)	76–111
Lymphatic filariasis	1,836 (106)	1,629–2,044	86 (9)	68–103
Schistosomiasis	1,811 (120)	1,576–2,046	101 (12)	78–125
Chagas disease	1,686 (151)	1,389–1,982	118 (17)	84–152
Leishmaniasis	1,022 (92)	841–1,203	52 (7)	38–66
Trachoma	649 (69)	514–784	20 (4)	13–28
Foodborne trematode–associated infections	610 (60)	492–729	41 (7)	28–54
Onchocerciasis	380 (47)	287–473	29 (12)	5–53
Yaws	161 (28)	106–216	7 (2)	3–11

*National estimates were determined on the Nationwide Inpatient Sample by using diagnostic codes from the International Classification of Diseases, 9th Revision, Clinical Modification. A complete list of ICD-9-CM codes used in this study is provided in the online Technical Appendix (wwwnc.cdc.gov/EID/article/21/6/14-1324-Techapp1.pdf)

Source: O'Neal SE, Flecker RH. Hospitalization frequency and charges for neurocysticercosis, United States, 2003–2012. Emerg. Infect. Dis. 21:969–976, 2015. Available at http://wwwnc.cdc.gov/eid/article/21/6/pdfs/14-1324.pdf. Accessed August 27, 2015.

order. The footnote provides supplementary information on how the data were obtained rather than repeating excessive methodological detail.

Note that these tables have three horizontal rules (lines) but no vertical rules. Virtually all tables are constructed this way. Occasionally, straddle rules (as below "Hospitalizations" and "Total charges" in Table 16.8) are used. Vertical rules normally are not used in tables.

EXPONENTS IN TABLE HEADINGS

If possible, avoid using exponents in table headings. Confusion has resulted because some journals use positive exponents and some use negative exponents to mean the same thing. For example, some have used "cpm $\times 10^{3}$" to refer to thousands of counts per minute, whereas others have used "cpm $\times 10^{-3}$" for the same thousands of counts. If it is not possible to avoid such labels in table headings (or in figures), it may be worthwhile to state in a footnote (or in the figure legend), in words that eliminate the ambiguity, what convention is being used.

FOLLOWING THE JOURNAL'S INSTRUCTIONS

Instructions to authors commonly include a section about tables. Before preparing your tables, check the instructions to authors of your target journal. These instructions may indicate such items as the dimensions of the space available, the symbols or form of lettering for indicating footnotes to tables, and the electronic tools to use in preparing tables. Looking at tables in the journal as examples also can aid in preparing suitable tables.

Style manuals in the sciences provide guidance in preparing not only text but also tables and figures. If your target journal specifies a style manual that it follows, consult it in this regard. Even if the journal does not specify a style manual, looking at one relevant to your field can aid in preparing effective tables and figures.

Traditionally, journals have asked authors to submit each table on a separate page at the end of the text. In addition, some journals have said to identify in the margin of the text the first mention of each table, for example, by writing "Table 3" and circling it. This procedure helps ensure that the author has indeed cited each table in the text, in numerical order. It also indicates to the compositor, at the page makeup stage, where to break the text to insert the tables. Today, some journals ask authors to embed tables in the text near their first mention. And some journals ask authors to submit tables as separate files. To determine whether tables should be placed within the text, placed at the end, or provided in separate files (and to determine how, if requested, to indicate their placement), consult the instructions to authors.

TITLES, FOOTNOTES, AND ABBREVIATIONS

The title of the table (or the legend of a figure) is like the title of the paper itself. That is, the title or legend should be concise and not divided into two or more clauses or sentences. Unnecessary words should be omitted.

Give careful thought to the footnotes to your tables. If abbreviations must be defined, you often can give all or most of the definitions in the first table. Then later tables can carry the simple footnote: "Abbreviations as in Table 1."

Note that "temp" (Tables 16.1, 16.2, 16.6, and 16.7) is used as an abbreviation for "temperature." Because of space limitations in tables, almost all journals encourage abbreviation of certain words in tables that would not be abbreviated in the text. Capitalize any such abbreviations used as the first word in a column heading; do not use periods (except after "no.," which might be misread without the period). To identify abbreviations that your target journal considers acceptable in tables, you can look at tables published in the journal. Also, some journals list in their instructions to authors the abbreviations that can be used without definition in tables that they publish.

ADDITIONAL TIPS ON TABLES

The following are some further tips to help ensure that you design and use tables effectively.

Use wording that will be clear without reference to the text. For example, a table should not just refer to "Group 1" and "Group 2." Rather, it should identify each group by a more meaningful designation (examples: "High-Dose Group" and "Low-Dose Group," "REM Sleep Group" and "NREM Sleep Group," and "Graduate Students" and "Professors").

If a paper includes a series of tables presenting analogous data, use an analogous format for each. For example, if several tables compare the same four groups from different standpoints, list the four groups in the same order in each table. Or if different tables present data on the same variables at different times, keep listing the variables in the same order. Such consistency saves readers effort. (And it is easier for you, too.)

Finally, remember to mention every table in the text. Do so as soon as readers are likely to want to see the table. You have gone to the effort of preparing good tables. Be sure that readers can benefit fully from them.

CHAPTER 17

How to Prepare Effective Graphs

A good illustration can help the scientist to be heard when speaking, to be read when writing. It can help in the sharing of information with other scientists. It can help to convince granting agencies to fund the research. It can help in the teaching of students. It can help to inform the public of the value of the work.

—Mary Helen Briscoe

WHEN NOT TO USE GRAPHS

In the previous chapter, we discussed certain types of data that should not be tabulated. They should not be turned into graphs either. Basically, graphs are pictorial tables.

The point is this. Certain types of data, particularly the sparse type or the type that is monotonously repetitive, do not need to be brought together in either a table or a graph. The facts are still the same: Preparing and printing an illustration can be time-consuming and expensive, and you should consider illustrating your data only if the result is a real service to the reader.

This point bears repeating because many authors, especially those who are still beginners, think that a table, graph, or chart somehow adds importance to the data. Thus, in the search for credibility, there is a tendency to convert a few data elements into an impressive-looking graph or table. Don't do it. Your more experienced peers and most journal editors will not be fooled; they will soon deduce that (for example) three of the four curves in your graph are simply the standard conditions and that the meaning of the fourth curve could have been stated in just a few words. Attempts to dress up scientific data are usually doomed to failure.

If there is only one curve on a proposed graph, can you describe it in words? Possibly only one value is really meaningful, either a maximum or a minimum; the rest is window dressing. If you determined, for example, that the optimum pH value for a particular reaction was pH 8.1, it would probably be sufficient to state something like, "Maximum yield was obtained at pH 8.1." If you determined that maximum growth of an organism occurred at 37°C, a simple statement to that effect is better economics and better science than a graph showing the same thing.

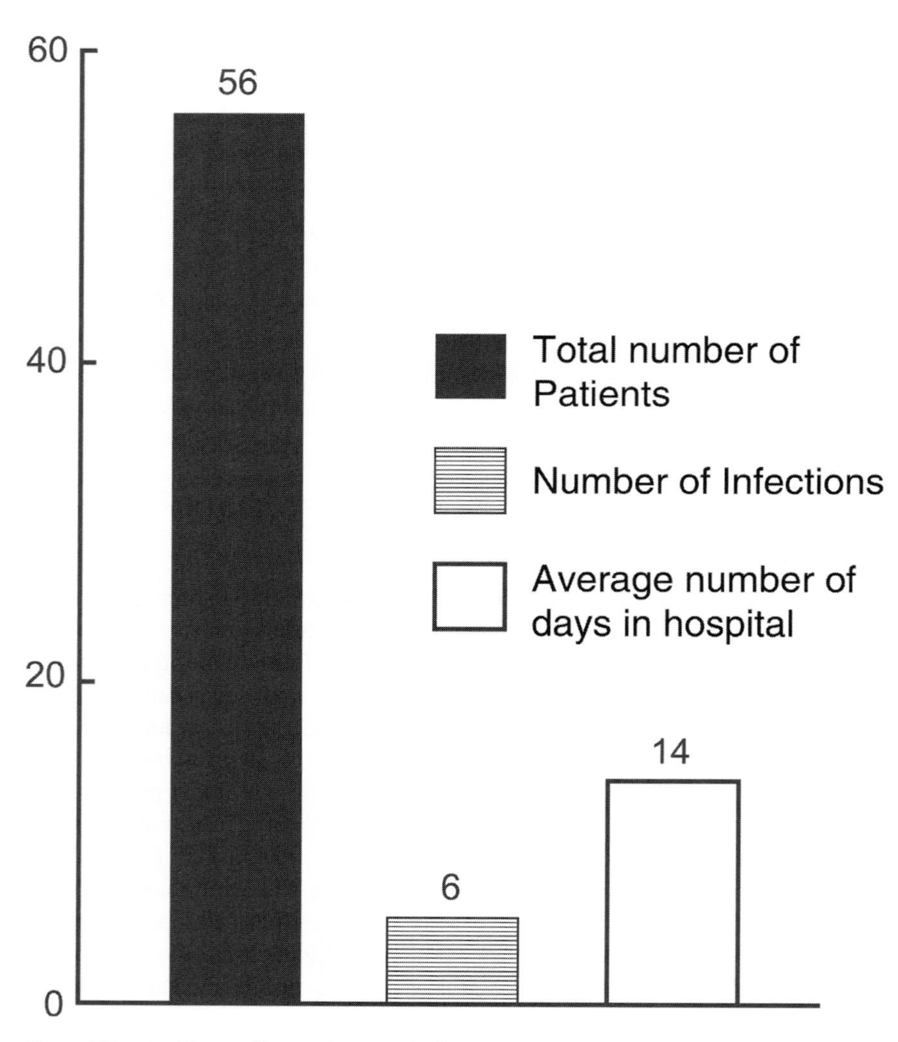

Figure 17.1. Incidence of hospital-acquired infections.

If the choice is not graph versus text but graph versus table, your choice might relate to whether you want to impart to readers exact numerical values or simply a picture of the trend or shape of the data. Rarely, there might be a reason to present the same data in both a table and a graph, the first presenting the exact values and the second showing a trend not otherwise apparent. Most editors would resist this obvious redundancy, however, unless the reason for it was compelling.

An example of an unneeded bar graph is shown in Fig. 17.1. This figure could be replaced by one sentence in the text: "Among the test group of 56 patients who were hospitalized for an average of 14 days, 6 acquired infections."

When is a graph justified? There are no clear rules, but let us examine some indications for their effective use.

WHEN TO USE GRAPHS

Graphs resemble tables as a means of presenting data in an organized way. In fact, the results of many experiments can be presented either as tables or as graphs. How do we decide which is preferable? This is often a difficult decision. A good rule might be this: If the data show pronounced trends, making an interesting picture, use a graph. If the numbers just sit there, with no exciting trend in evidence, a table should be satisfactory (and perhaps easier and cheaper for you to prepare). Tables are also preferred for presenting *exact* numbers.

Examine Table 17.1 and Fig. 17.2, both of which record exactly the same data. Either format would be acceptable for publication, but Fig. 17.2 clearly seems superior to Table 17.1. In the figure, the synergistic action of the two-drug combination is immediately apparent. Thus, the reader can quickly grasp the significance of the data. It also appears from the graph that streptomycin is more effective than is isoniazid, although its action is somewhat slower; this aspect of the results is not readily apparent from the table.

Table 17.1. Effect of streptomycin, isoniazid, and streptomycin plus isoniazid on *Mycobacterium tuberculosis*[a]

	Percentage of negative cultures at:			
Treatment[b]	2 wk	4 wk	6 wk	8 wk
Streptomycin	5	10	15	20
Isoniazid	8	12	15	15
Streptomycin + Isoniazid	30	60	80	100

[a]The patient population, now somewhat less so, was described in a preceding paper (61).
[b]Highest quality available from our supplier (Town Pharmacy, Podunk, IA).

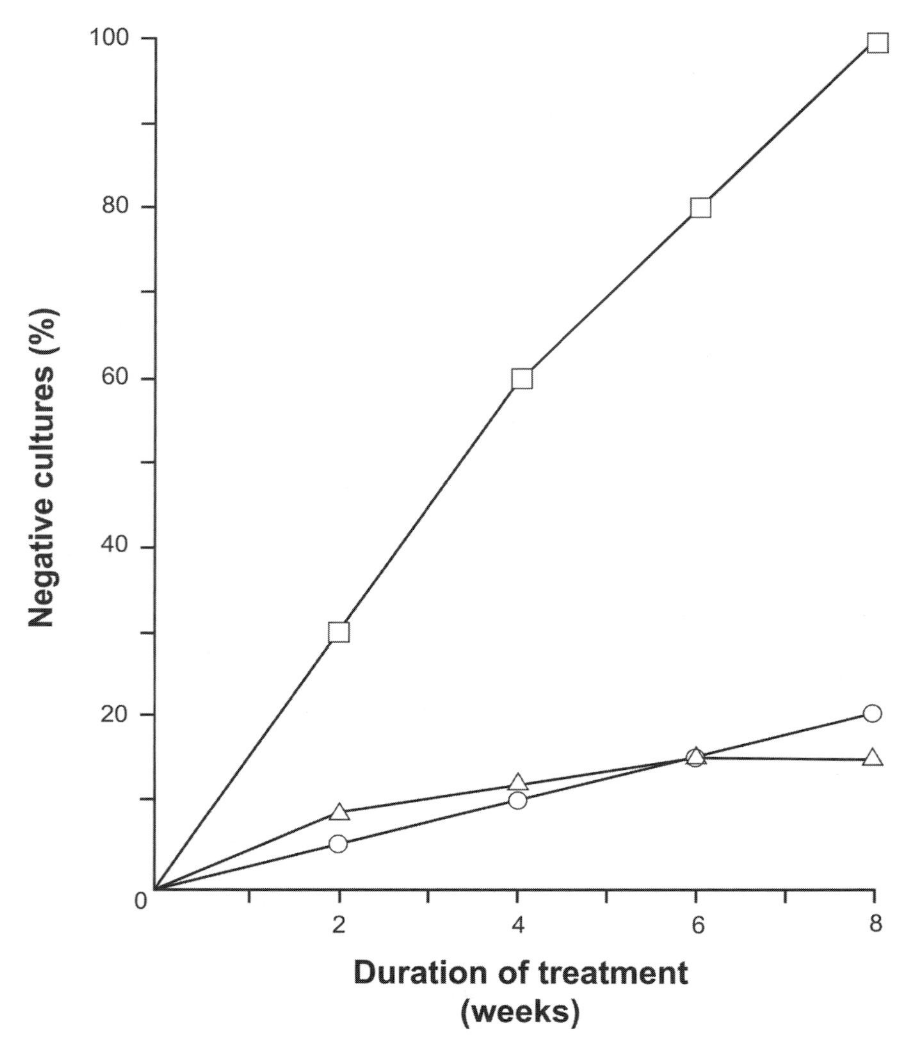

Figure 17.2. Effect of streptomycin (○), isoniazid (△), and streptomycin plus isoniazid (□) on *Mycobacterium tuberculosis.*

HOW TO PREPARE GRAPHS

Early editions of this book included rather precise directions for using graph paper, India ink, lettering sets, and the like. Graphs had been prepared with these materials and by these techniques for generations.

Today we prepare graphs by computer. However, the *principles* of producing good graphs have not changed. The sizes of the letters and symbols, for example, must be chosen so that the final published graph in the journal is clear and readable.

The size of the lettering must be based on the anticipated reduction that will occur in the publishing process. This factor can be especially important if you are combining two or more graphs into a single illustration. Remember: Text that is easy to read on a large computer screen may become illegible when reduced to the width of a journal column.

Each graph should be as simple as possible. "The most common disaster in illustrating is to include too much information in one figure. Too much information in an illustration confuses and discourages the viewer" (Briscoe 1996).

Figure 17.3 is a nice graph. The lettering is large enough to read easily. It is boxed, rather than two-sided (compare with Figure 17.2), making it a bit easier to estimate the values on the right-hand side of the graph. The scribe marks point inward rather than outward.

If your paper contains two or more graphs that are most meaningful when viewed together, consider grouping them in a single illustration. To maximize readability, place the graphs above and below each other rather than side by side. For example, in a two-column journal, placing three graphs in an "above and below" arrangement allows each graph to be one or two columns in width. If the graphs appear side by side, each can average only one third of a page wide.

Whether or not you group graphs in such a composite arrangement, be consistent from graph to graph. For example, if you are comparing interventions, keep using the same symbol for the same intervention. Also be consistent in other aspects of design. Both conceptually and aesthetically, the graphs in your paper should function as a set.

Do not extend the ordinate or the abscissa (or the explanatory wording) beyond what the graph demands. For example, if your data points range between 0 and 78, your topmost index number should be 80. You might feel a tendency to extend the graph to 100, a nice round number; this urge is especially difficult to resist if the data points are percentages, for which the natural range is 0 to 100. Resist this urge, however. If you do not, parts of your graph will be empty; worse, the live part of your graph will then be restricted in dimension, because you have wasted perhaps 20 percent or more of the width (or height) with empty white space.

In the preceding example (data points ranging from 0 to 78), your reference numbers should be 0, 20, 40, 60, and 80. You should use short index lines at each of these numbers and also at the intermediate 10s (10, 30, 50, 70). Obviously, a reference stub line halfway between 0 and 20 could only be 10. Thus, you need not letter the 10s, and you can then use larger lettering for the 20s, without

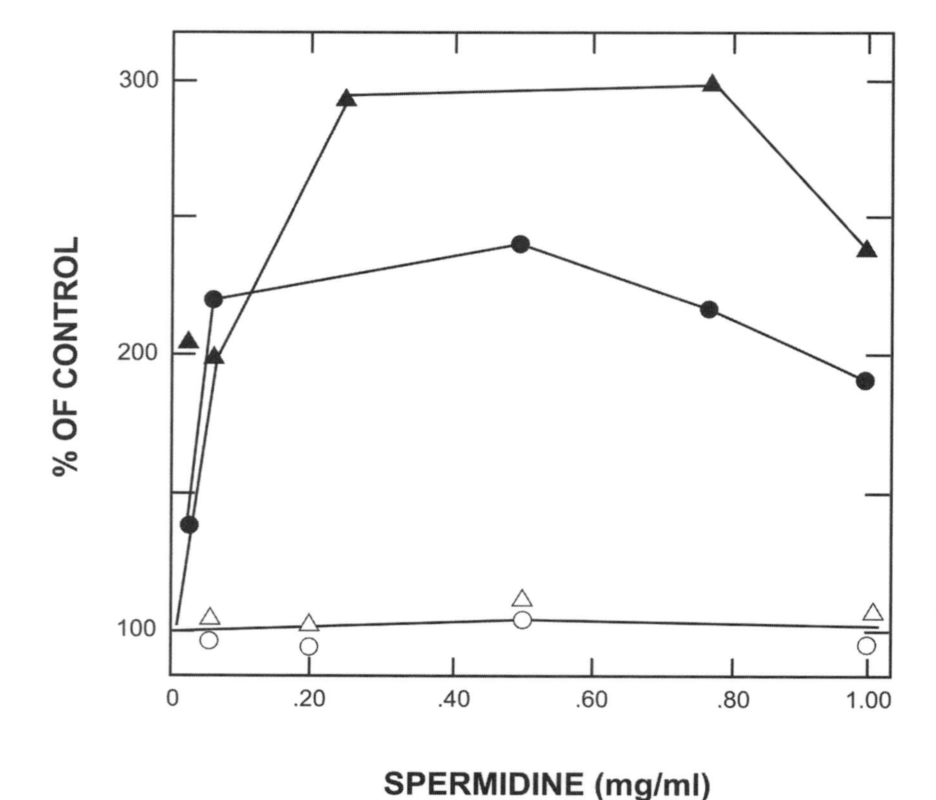

Figure 17.3. Effect of spermidine on the transformation of *B. subtilis* BR 151. Competent cells were incubated for 40 min with spermidine prior to the addition of 5 μg of donor DNA per ml (●) or 0.5 μg of donor DNA per ml (▲). DNA samples of 5 μg (○) or 0.5 μg per ml (△) were incubated for 20 min prior to the addition of cells. (Redrawn from Fig. 1 in Clark PO and Leach FR. Stimulation of *Bacillus subtilis* transformation by spermidine. Mol. Gen. Genet. 178:21–25, 1980. © by Springer-Verlag 1980. With permission of Springer.)

squeezing. By using such techniques, you can make graphs simple and effective instead of cluttered and confusing.

SYMBOLS AND LEGENDS

If there is a space in the graph itself, use it to present the key to the symbols. In the bar graph (Figure 17.1), the shadings of the bars would have been a bit difficult to define in the legend; given as a key, they need no further definition (and any additional typesetting, proofreading, and expense are avoided).

If you must define the symbols in the figure legend, you should use only those symbols that are considered standard and that are widely available. Perhaps the most standard symbols are open and closed circles, triangles, and squares (○, △, □, ●, ▲, ■). If you have just one curve, use open circles for the reference points; use open triangles for the second, open squares for the third, closed circles for the fourth, and so on. If you need more symbols, you probably have too many curves for one graph, and you should consider dividing it into two. Different types of connecting lines (solid, dashed) can also be used. But do *not* use different types of connecting lines *and* different symbols.

As to the legends, they should *normally* be provided on a separate page, not at the bottom or top of the illustrations themselves. The main reason is that the two portions commonly are processed separately during journal production. Consult the instructions to authors of your target journal regarding this matter and other requirements for graphs.

A FEW MORE TIPS ON GRAPHS

Design graphs, like tables, to be understandable without the text. For example, use meaningful designations (not just numbers) to identify groups. And refer to each graph as soon as readers are likely to want to see it. Do not leave readers trying to visualize your findings by sketching them on a napkin—only to find three pages later that a graph displays them.

Use graphs that depict your findings fairly and accurately. For example, do not adapt the scales on the axes to make your findings seem more striking than they are. With rare exceptions, avoid beginning a scale at anything other than zero. And if you interrupt a scale line to condense a graph, make the interruption obvious. Also, if the standard deviation is the appropriate way to show the variability in your data, do not substitute the standard error of the mean, which might make the data seem more consistent than it is.

Note that some journals (mainly the larger and wealthier ones) redraw graphs and some other types of figures to suit their own format. Whether or not a journal will do so, prepare your graphs well. Doing so will help make your findings and their value clear and will help show the care with which you do your work.

CHAPTER 18

How to Prepare Effective Photographs

Life is not about significant details, fixed in a flash, fixed forever. Photographs are.

—Susan Sontag

PHOTOGRAPHS AND MICROGRAPHS

If your paper is to be illustrated with one or more photographs, there are several factors to keep in mind.

The most important factor to worry about, however, is a proper appreciation of the *value* of the photographs for the story you are presenting. The value can range from essentially zero (in which case, like useless tables and graphs, they should not be submitted) to a value that transcends that of the text itself. In many studies of cell ultrastructure, for example, the significance of the paper lies in the photographs. If photographs (such as electron micrographs) are of prime importance to your story, ensure the journal you choose has high-quality reproduction standards, as discussed in Chapter 6.

As with graphs, the size (especially width) of the photograph in relation to the column and page width of the journal can be important. Try to avoid dimensions that will require excessive reduction of the photograph to suit the journal page.

SUBMISSION FORMATS

Today, journals normally request photographs in electronic format. To ascertain requirements for photographs, see the instructions to authors for your target journal. For example, check what formats (such as EPS, JPEG, or TIFF) are acceptable and what resolution is required. As indicated in Chapter 5, care must be taken to avoid making unwarranted changes in digital photographs. Sources of guidance on using digital images ethically include writings by Cromey (2010, 2012).

CROPPING

Whatever the quality of your photographs, you want to have them published legibly. To some degree, you can control this process yourself if you use your head.

If you are concerned that detail might be lost by excessive reduction, there are several ways you might avoid this. Seldom do you need the whole photograph, right out to all four edges. Therefore, crop the photograph to include only the important part. Commonly, photographs are cropped digitally. If you are submitting a print, you can write "crop marks" on the margin to show where the photograph should be cropped. Figures 18.1 and 18.2 show photographs with and without cropping.

NECESSARY KEYS AND GUIDES

If you can't crop down to the features of special interest, consider superimposing arrows or letters on the photographs, as shown in Figure 18.3. In this way, you can draw the reader's attention to the significant features. Having arrows or letters to refer to can aid in writing clear, concise legends.

Unless your journal requests that photographs and other illustrations be embedded in the text, it is a good idea to indicate the preferred location for each illustration. In this way, you will be sure that all illustrations have been referred to in the text, in one-two-three order, and the printer will know how to weave the illustrations into the text so that each one is close to the text related to it.

With electron micrographs, put a micrometer marker directly on the micrograph. In this way, regardless of any reduction (or even enlargement) in the printing process, the magnification factor is clearly evident. The practice of putting the magnification in the legend (for example, $\times 50{,}000$) is not advisable, and some journals no longer allow it, precisely because the size (and thus magnification) is likely to change in printing. And, usually, the author forgets to change the magnification at the proof stage.

In other photographs where the size of the object is important, likewise include a scale bar. Sometimes showing a familiar object, such as a paper clip, near the object can help readers discern an object's size. Remember, though, that some objects (such as coins of given denominations) that are familiar to readers in one country might be unfamiliar to readers elsewhere.

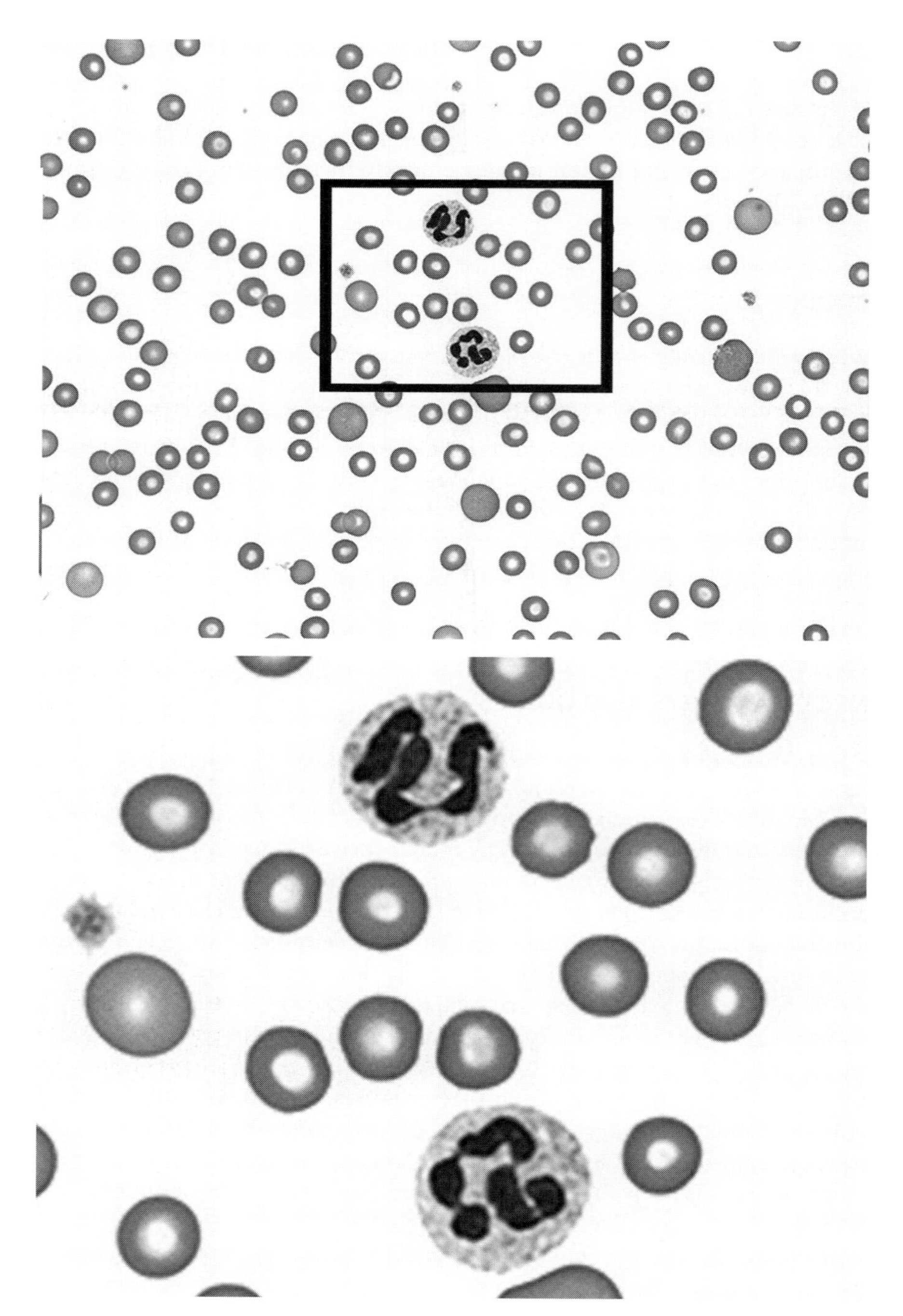

Figure 18.1. Uncropped (top) and cropped versions of a photograph. (Courtesy of CVM Communications, College of Veterinary Medicine and Biomedical Sciences, Texas A&M University.)

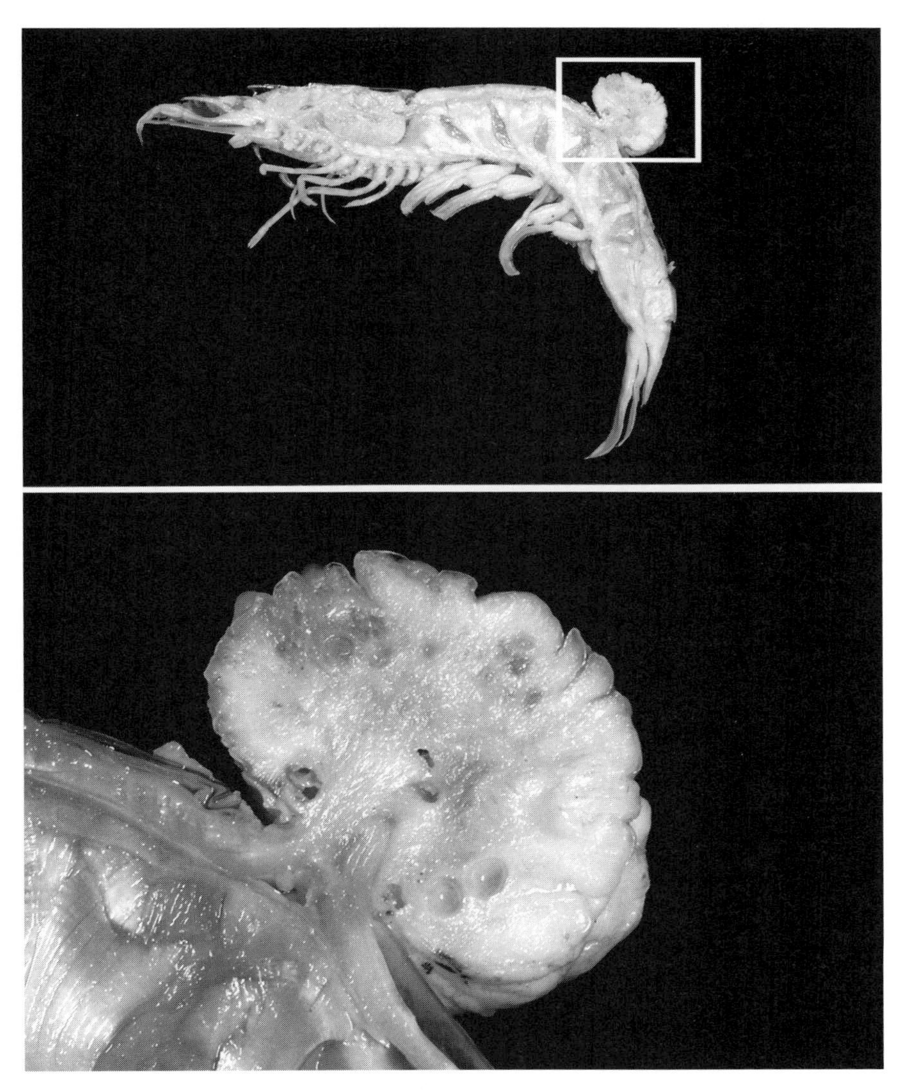

Figure 18.2. Uncropped (top) and cropped versions of a photograph. In this case, publication of both versions may be warranted, to both show the tumor in context and display detail. (Courtesy of CVM Communications, College of Veterinary Medicine and Biomedical Sciences, Texas A&M University.)

COLOR

Until recently, journals seldom published color photographs and other color illustrations, because of the high cost of printing them. Today, however, color printing has become more affordable. And for articles online, color does not

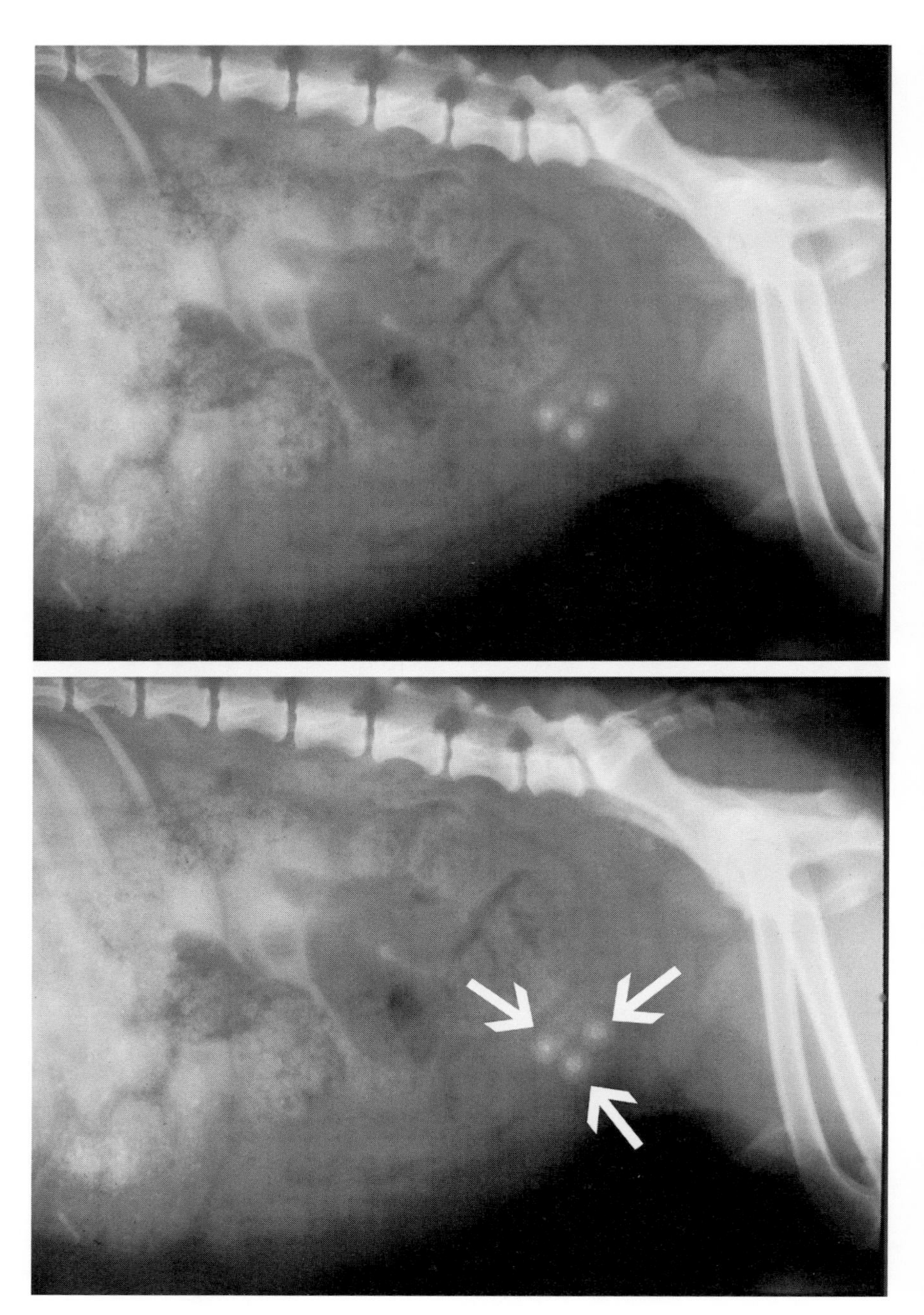

Figure 18.3. An example of adding arrows to direct readers' attention to structures of interest. (Courtesy of CVM Communications, College of Veterinary Medicine and Biomedical Sciences, Texas A&M University.)

increase cost. Thus, more color illustrations are appearing, and use of color has become relatively common in some fields and journals. If you have the option of including color, consider whether doing so will improve your scientific paper. Would color help to tell your story? Or would it be merely decorative or even distracting?

If you are considering using color, see the instructions to authors of your target journal for specifications regarding color illustrations and for information on any charges for color. If color illustrations are to be printed, authors commonly must pay a fee. Some journals, however, do not charge for color. For example, the guidelines for illustrations in American Chemical Society journals state: "The use of color to enhance the clarity of complex structures, figures, spectra, schemes, etc. is encouraged. Color reproduction of graphics will be provided at no cost to the author."

LINE DRAWINGS

In some fields (for example, descriptive biology), line drawings are superior to photographs in showing important details. Such illustrations are also common in medicine, especially in presenting anatomic views, and indeed have become virtually an art form. When illustrations are necessary, the services of a professional illustrator generally are required. Such illustrators are available at many universities and other research institutions and can be identified through associations of scientific and medical illustrators.

PART IV

Publishing the Paper

CHAPTER 19

Rights and Permissions

Take away from English authors their copyrights, and you would very soon take away from England her authors.

—Anthony Trollope

WHAT IS COPYRIGHT?

Before you submit your paper to a journal, you should be aware of two items regarding copyright. First, if your paper includes illustrations or other materials that have been published elsewhere, you will need permission to republish them unless you hold the copyright. Second, you may need to transfer the copyright for your paper to the journal (or, for some journals, transfer limited rights while retaining copyright).

Copyright is the exclusive legal right to reproduce, publish, and sell the matter and form of a literary or artistic work. (Here "literary and artistic" is broadly defined and so includes scientific papers.) Copyright protects original forms of expression but not the ideas being expressed. The data you are presenting are not protected by copyright; however, the collection of the data and the way you have presented them are protected. You own the copyright of a paper you wrote for the length of your life plus 50 years, as long as it was not done for an employer or commissioned as work for hire. If you have collaborated on the work, each person is a co-owner of the copyright, with equal rights.

Copyright is divisible. The owner of the copyright may grant one person a nonexclusive right to reproduce the work and another the right to prepare derivative works based on the copyrighted work. Copyright can also be transferred. Transfers of the copyright must be made in writing by the owner. An employer may transfer copyright to the individual who developed the original work. If you wish to copy, reprint, or republish all or portions of a copyrighted work that you do not own, you must get permission from the copyright owner. If you, as an author, have transferred the complete copyright of your work to a publisher, you must obtain permission for use of your own material from the publisher.

Fair use of copyrighted material, according to the 1976 Copyright Act, allows you to copy and distribute small sections of a copyrighted work. It does not allow you to copy complete articles and republish them without permission, whether for profit or otherwise.

COPYRIGHT CONSIDERATIONS

The legal reasons for seeking permission when republishing someone else's work relate to copyright law. If a journal is copyrighted, and most of them are, legal ownership of the published papers becomes vested in the copyright holder. Thus, if you wish to republish copyrighted material, you must obtain approval of the copyright holder or risk suit for infringement.

Publishers acquire copyright so that they will have the legal basis, acting in their own interests and on behalf of all authors whose work is contained in the journals, for preventing unauthorized use of such published work. Thus, the publishing company and its authors are protected against plagiarism, misappropriation of published data, unauthorized reprinting for advertising and other purposes, and other potential misuse.

In the United States, under the 1909 Assignment of Copyright Act, submission of a manuscript to a journal was presumed to carry with it assignment of the author's ownership to the journal (publisher). Upon publication of the journal, with the appropriate copyright imprint in place and followed by the filing of copies and necessary fees with the Register of Copyrights, ownership of all articles contained in the issue effectively passed from the authors to the publisher.

The Copyright Act of 1976 requires that henceforth this assignment may no longer be assumed; it must be in writing. In the absence of a written transfer of copyright, the publisher is presumed to have acquired only the privilege of publishing the article in the journal itself; the publisher would then lack the right to produce reprints, copies, and electronic forms or to license others to do so (or to legally prevent others from doing so). Also, the Copyright Act stated that copyright protection begins "when the pen leaves the paper" (equivalent today to "when the fingers leave the keyboard"), thus recognizing the intellectual property rights of authors as being distinct from the process of publication.

Therefore, most publishers now require that each author contributing to a journal assign copyright to the publisher, either when the manuscript is submitted or when it is accepted for publication. To effect this assignment, the publisher provides each submitting author with a document usually titled "Copyright Transfer Form." Such forms are commonly available on the websites of journals.

Another feature of the 1976 Copyright Act that may interest scientists, both as authors and as users of the research literature, deals with copying. On the one hand, authors wish to see their papers receive wide distribution. On the other hand, they may not want this to take place at the expense of the journal. Thus, the law reflects these conflicting interests by defining certain kinds of library and educational copying as "fair use" (that is, copying that may be done without permission and without payment of royalties), while at the same time protecting the publisher against unauthorized systematic copying.

To make it easy to authorize systematic copiers to use journal articles and to remit royalties to publishers, a Copyright Clearance Center (www.copyright.com) has been established. Most scientific publishers of substantial size have joined the center. This central clearinghouse makes it possible for a user to make as many copies as desired, without needing to obtain prior permission, if the user is willing to pay the publisher's stated royalty to the center. Thus, the user need deal with only one source, rather than facing the necessity of getting permission from and then paying royalties to hundreds of different publishers.

Because both scientific ethics and copyright law are of fundamental importance, every scientist must be acutely sensitive to them. Basically, this means that you must not republish tables, figures, and substantial portions of text *unless* you have acquired permission from the owner of the copyright. Even then, it is important that you label such reprinted materials, usually with a credit line reading "Reprinted with permission from (journal or book reference); copyright (year) by (owner of copyright)." Often, information on how to seek permission is posted on the website of the journal or other publication in which the material appeared. If the website does not provide the information, contact the editorial office of the publication.

COPYRIGHT AND ELECTRONIC PUBLISHING

Traditionally, journals and books have been well defined as legal entities. However, once the same information enters a digital environment, it becomes a compound document that includes not only text but also programming code and database access information that has usually been created by someone (often several people) other than the author of the paper. All copyright law, and all rules and regulations pertaining to copyright, hold true for electronic publication, including material posted on the Internet. Unless the author or owner of the copyright of work posted on the Internet has placed on that work a specific note stating that the item is in the public domain, it is under copyright and you may not reproduce it without permission. Although you need not post a copyright notice for protection of your Internet materials, doing so acts as a warning to people who might use your material without permission. To post such a notice, you need only place the word "Copyright," the date of the publication, and the name of the author or copyright owner near the title of the work, for example, "Copyright 2015 by Magon Thompson (or Sundown Press)."

The electronic era has brought with it an interest in alternatives to transfer of copyright—in particular, the use of licenses, such as those developed by Creative Commons (creativecommons.org), allowing limited rights to works. Some open access journals use this approach, in which the authors retain copyright but allow reproduction of their work under specified conditions, such as attributing the work to the authors. If you publish in a journal using such licenses, you will be asked to complete such an agreement rather than a copyright transfer form. Information about various journal publishers' copyright agreements can be accessed at the website SHERPA/RoMEO (www.sherpa.ac.uk/romeo/).

As electronic publishing evolves further, additional developments relating to copyright and permissions may well occur. Whether seeking to include material published elsewhere or seeking to publish your own work, look for the latest word from the publishers involved.

CHAPTER 20

How to Submit the Manuscript

Great journals are born in the hands of the editors; they die in the hands of businessmen.

—Bernard DeVoto

CHECKING YOUR MANUSCRIPT

Before submitting your manuscript, review the instructions to authors from the journal. If the journal provides a manuscript-submission checklist, remember to use it. Make sure you have followed all instructions. If a manuscript deviates substantially from what is required, it may be returned for correction of the problems before it undergoes review.

Unless the journal (or the style manual that it says to use) instructs otherwise, follow these guidelines:

- Double-space.
- Use margins of at least 1 inch (at least about 25 mm).
- Left-justify the text; leave a "ragged" right margin.
- Start each section of the manuscript on a new page. The title and authors' names and addresses are usually on the first page, which should be numbered 1. The abstract is on the second page. The introduction starts on the third page, and each succeeding section (materials and methods, results, etc.) then starts on a fresh page. Figure legends are grouped on a separate page. Traditionally, the tables, figures, and figure legends have been assembled at the back of the manuscript. Recently, though, some journals have asked authors to insert them in the text or to provide them as separate files.

Grammar-checking and spell-checking functions can help but should not be relied on too heavily. Grammar checkers associated with word-processing programs can alert you to possible problems in grammar and style. But given their limitations, you should accept their suggestions only if you confirm that they are correct. Almost all spell-checkers provide for the creation of custom dictionaries—for example, for scientific terms and unusual words; also, some specialized spell-checkers are commercially available. Spell-checkers recognize definite misspellings but not those typographical errors that result in the wrong word (for example, as occurs embarrassingly often, "pubic" instead of "public"). Thus, proofreading still is necessary, to make sure the correct word has appeared and to detect errors such as missing words. In addition to proofreading the manuscript yourself, try to have someone do so who has not seen the manuscript before and therefore may notice problems that you miss. Consider also reading the manuscript aloud, as doing so can aid in noticing difficulties.

SUBMITTING YOUR MANUSCRIPT

For most of the history of scientific publication, researchers submitted their papers by mail. Yes, physical mail. Yes, the postal service, not email. In fact, earlier editions of this book advised readers on how to package a manuscript for mailing and what class of mail to use. Today, if a journal asks you to submit a hard copy of your manuscript by post, take a clue: The journal probably still is still stuck in the last century—which presumably is not where you want the output of your new research to be.

Today, journals commonly have online submission systems through which authors must submit their manuscripts. Journals that are small or are not associated with major publishers sometimes request manuscripts simply as email attachments. Over the years, online systems for electronic submission have become relatively easy to use. And conveniently, they can allow authors to track the progress of their papers toward publication. For guidance on how to submit your manuscript, consult the journal's instructions to authors and look at the journal's website.

THE COVER LETTER

Finally, you should submit a cover letter with the manuscript (or provide equivalent information if an online submission system prompts you to do so). This letter, which is from the corresponding author, provides context for considering your paper. Most basically, it identifies the title of the article, the authors, and

the journal to which the paper is being submitted. It may also identify the type of submission (for example, scientific paper or review article) and, if applicable, indicate the intended subject-matter section of the journal. Commonly, the letter must attest that the work is original and that the manuscript is not being considered by other journals. (Whereas one may apply to multiple graduate schools for admission and accept the best offer, standard practice is to submit a paper to one journal at a time. If the paper is not accepted, it can then be submitted to another journal.)

The letter also may attest that all the listed authors qualify to be listed and that no one meeting the criteria for authors was excluded. In addition, it may indicate whether the authors have conflicts of interest and, if so, what these conflicts are. If some content of the paper has appeared previously, for example in a conference abstract, the cover letter typically should state so. The letter also may do other items, such as mention a photograph or other image in the paper that may be well suited to appear on the cover of the journal.

Sometimes cover letters also suggest potential peer reviewers of the paper—or identify individuals whom the authors do not want as peer reviewers. Suggested reviewers should be scientists who can review the paper knowledgeably and without bias. They should not be people with conflicts of interest (for example, colleagues at the authors' institution or mentors, close friends, or family members of the authors). They can, however—as may be inevitable in small fields—be people whom the authors have met at conferences or otherwise know casually. Only if serious reason exists to believe that someone will be biased or otherwise unsuitable should a request be made to exclude a potential reviewer. Such a request may be made if, for example, someone in the authors' research area has had a major professional conflict with an author, is suspected of unscrupulous behavior, or is a spouse—or former spouse—of an author. Suggestions regarding reviewers are simply that: suggestions. The journal may follow them or not; commonly, they may use some reviewers suggested by authors plus some other reviewers. Journals often appreciate, and sometimes request or require, names and contact information of potential reviewers.

Explaining why the paper is believed to merit publication in the journal can also be useful in the cover letter. Doing so may be especially helpful if the relevance of the subject matter to the journal or the novelty or value of the research may not be immediately apparent. It also may be especially valuable if the journal is of broad scope, in which case the editor who first sees the paper might not be very familiar with the research topic and so might not readily recognize the importance of the contribution.

Some journals' instructions to authors include guidance on what to include in the cover letter. If so, of course proceed accordingly.

SAMPLE COVER LETTER

Dear Dr. _________:

Accompanying this letter is a manuscript by Mary Q. Smith and Adam B. Appiah titled "Fatty Acid Metabolism in *Cedecia neteri*," which is being submitted for possible publication in the Physiology and Metabolism section of the *Journal of Bacteriology*.

This manuscript is new, is not being considered elsewhere, and reports new findings that extend results we reported earlier in the *Journal of*

Biological Chemistry (284:112–117, 2014). An abstract of this manuscript was presented earlier (Abstr. Annu. Meet. Am. Soc. Microbiol., p. 406, 2015).

Sincerely,
Mary Q. Smith

The above is a model of a brief, simple cover letter. A more extensive example appears at journals.lww.com/greenjournal/Documents/SampleCoverLetter.pdf. You may find this letter, which introduces the (fictional) paper "Primary Cesarean Delivery Among Pandas," both instructive and amusing.

ELECTRONIC COVER LETTERS

If you are submitting your manuscript electronically, the manuscript-submission website may supply a mechanism for providing your cover letter. Alternatively, it may prompt you for the information the journal wants to receive, thus automatically generating a cover letter or the equivalent. This electronic option saves you the trouble of composing a letter and helps ensure that the journal receives the information it requires.

CONFIRMATION OF RECEIPT

Most journals send out an "acknowledgment of receipt" by email or other means when the manuscript is received or have a mechanism by which authors check the journal website to see whether submission is complete. If you do not receive acknowledgment within 2 weeks (or less for electronically submitted manuscripts), call or write to the editorial office to verify that your manuscript was indeed received. We know of one author whose manuscript was lost in transit, and not until 9 months later was the problem brought to light by his meek inquiry as to whether the reviewers had reached a decision about his manuscript. Do be sure that your manuscript was received.

CHAPTER 21

The Review Process (How to Deal with Editors)

Many editors see themselves as gifted sculptors, attempting to turn a block of marble into a lovely statue, and writers as crude chisels. In actual fact, the writers are the statues, and the editors are pigeons.

—Doug Robarchek

FUNCTIONS OF EDITORS, MANAGING EDITORS, AND MANUSCRIPT EDITORS

Editors and managing editors have impossible jobs. What makes their work impossible is the attitude of authors. This attitude was well expressed by Earl H. Wood of the Mayo Clinic in his contribution to a panel on the subject "What the Author Expects from the Editor." Dr. Wood said, "I expect the editor to accept all my papers, accept them as they are submitted, and publish them promptly. I also expect him to scrutinize all other papers with utmost care, especially those of my competitors."

Somebody once said, "Editors are, in my opinion, a low form of life—inferior to the viruses and only slightly above academic deans."

And then there is the story about the Pope and the editor, both of whom died and arrived in heaven simultaneously. They were subjected to the usual initial processing and then assigned to their heavenly quarters. The Pope looked around his apartment and found it to be spartan indeed. The editor, on the other hand, was assigned to a magnificent apartment, with plush furniture, deep-pile carpets, and superb appointments. When the Pope saw this, he went to God and said: "Perhaps there has been a mistake. I am the Pope and I have been assigned to shabby quarters, whereas this lowly editor has been assigned to a lovely apartment." God answered: "Well, in my opinion, there isn't anything very special about you. We've admitted 200 Popes in the last 2,000 years. But this is the very first editor who ever made it to heaven."

Going back to the first sentence in this chapter, let us distinguish between editors and managing editors. Authors should know the difference, if for no other reason than knowing to whom to complain to when things go wrong.

An *editor* (some journals have several) decides whether to accept or reject manuscripts. Thus, the editor of a scientific journal is a scientist, often of preeminent standing. The editor not only makes the final "accept" and "reject" decisions but also designates the peer reviewers upon whom he or she relies for advice. When you have reason to object to the quality of the reviews of your paper (or the decision reached), your complaint should be directed to the editor. (Adlai Stevenson joked that the role of the editor is to separate the wheat from the chaff and then make sure that the chaff gets printed.)

Especially at larger journals, there may be several such editors. For example, there may be an *editor in chief* (the top editor, in charge of overall journal content), a second in command known as a *deputy editor,* and a few associate or assistant editors. Sometimes different associate or assistant editors oversee the review of papers in different subject areas covered by the journal. Collectively, the editor in chief and other editors involved in evaluating and choosing papers sometimes are called *scientific editors.*

The *managing editor* is normally a full-time paid professional, whereas editors commonly are unpaid volunteer scientists. (A few large scientific and medical journals do have full-time paid editors. Some other journals, especially those that are in medical fields or are published commercially, pay salaries to their part-time editors.) Normally, the managing editor is not directly involved with the accept-reject decisions. Instead, the managing editor attempts to relieve the editor of all clerical and administrative detail in the review process, and he or she is responsible for the later events that convert accepted manuscripts into published papers. Thus, when problems occur at the proof and publication stages, you should communicate with the managing editor.

In short, preacceptance problems are normally within the province of the editor, whereas postacceptance problems are within the bailiwick of the managing editor. However, managing editors have observed that there seems to be one fundamental law that everybody subscribes to: "Whenever anything goes wrong, blame the managing editor."

Another editor you may encounter once your paper is accepted is a *manuscript editor,* also known as a *copy editor.* This individual may be a staff member working at the journal office or publishing company or a freelance contractor working at home in pajamas. The manuscript editor edits your paper for consistency with the journal style and format. In addition, he or she corrects errors in grammar, spelling, punctuation, and usage. At some journals, he or she also works to improve expression in other ways, for example by making wording clearer and more concise. If the manuscript editor has questions (for instance, about inconsistencies between numbers in a table and in the text), he or she will ask the author for clarification, by submitting what are called *queries.* View the manuscript editor as an ally in communicating your research to your readers and presenting yourself well to your professional community. Or, as one author told a manuscript editor, "Until I saw your edited version of my paper, I didn't realize how brilliant I was."

THE REVIEW PROCESS

You, as an author, should have some idea of the whys and wherefores of the review process. Therefore, we will describe the policies and procedures that are typical in most editorial offices. If you understand (and perhaps even appreciate) some of the reasons for the editorial decisions that are made, perhaps you can facilitate publication of your papers simply by knowing how to deal with editors.

When your manuscript arrives at the journal editorial office, the editor (or the managing editor, if the journal has one) makes several preliminary decisions. First, is the manuscript concerned with a subject area covered by the scope of the journal? If it clearly is not, the manuscript is immediately returned to the submitting author, along with a short statement of the reason for the action. Seldom would an author be able to challenge such a decision successfully, and it is usually pointless to try. It is an important part of the editor's job to define the scope of the journal, and editors seldom take kindly to suggestions by authors, no matter how politely the comments are phrased, that the editor is somehow incapable of defining the basic character of his or her journal. Remember, however, that such a decision is not rejection of your data or conclusions. Your course of action is obvious: Try another journal.

Second, if the subject of the manuscript is appropriate for consideration, is the manuscript itself in suitable form for consideration? Is the manuscript complete, with no sections, tables, or figures missing? Is the manuscript in the editorial style of the journal, at least as to the basics? If the answer to either of the preceding questions is no, the manuscript may be immediately returned to the author or, at the least, the review will be delayed while the deficiencies are rectified. Most journal editors will not waste the time of their valued editorial board members and consultants by sending poorly prepared manuscripts to them for review.

One editor, a kindly man by nature, became exasperated when a poorly prepared manuscript that had been returned to the author was resubmitted to the journal with very little change. The editor then wrote the following letter, which is printed here as a warning to all students of the sciences everywhere:

> Dear Dr. ____________:
>
> I refer to your manuscript ____________ and have noted in your letter of August 23 that you apologize without excuse for the condition of the original submission. There is really no excuse for the rubbish that you have sent forward in the resubmission.
>
> The manuscript is herewith returned to you. We suggest that you find another journal.
>
> Yours sincerely,
>
> ________________

Only after these two preconditions (a proper manuscript on a proper subject) have been met is the editor ready to consider the manuscript for publication.

At this point, the editor must perform two very important functions. First, the basic housekeeping must be done. That is, careful records should be established so that the manuscript can be followed throughout the review process and (if the manuscript is accepted) the publication process. If the journal has a managing editor, and most of the large ones do, this activity is normally a part of his or her assignment. It is important that this work be done accurately, so that the whereabouts of manuscripts are known at all times. It is also important that the system include a number of built-in signaling devices, so that the inevitable delays in review and other problems can promptly be brought to the attention of the editor or managing editor. The electronic systems that many journals use for manuscript submission and tracking facilitate this work.

Second, the editor must decide whether the paper will be peer reviewed (evaluated by other experts in the same research field) and, if so, choose peer reviewers. At many journals, all manuscripts reaching this stage are sent for peer review. At some journals—especially the larger and more competitive ones, which receive very many papers—the editors decide which manuscripts will be peer reviewed. If the editors know they would not publish the paper, for example because the research is too weak or the topic is too narrow, they return the paper to the author without peer review. Such return generally is quick; thus, the author does not waste weeks or more awaiting the unfavorable decision, as could well occur if the paper went for peer review. If you receive a rapid rejection—sometimes known as a "desk rejection" or "desk reject"—realize that you are not alone, and submit your paper to another, perhaps more specialized journal. Of course, carefully consult the new journal's instructions to authors first.

If the paper will be sent for peer review—as probably is the case for most papers at most journals—the editor or editors must choose the peer reviewers (also known as referees). Commonly, two reviewers are selected for each manuscript; in some fields, however, three or more reviewers often are used, especially for interdisciplinary papers, and in some fields, use of a single reviewer is the norm. The reviewers must be peers of the author—that is, fellow experts—or their recommendations will be of little value. Frequently, the editor starts with the editorial board of the journal. Who on the board has the appropriate subject expertise to evaluate a particular manuscript? Often, because of the highly specialized character of modern science, only one member (or no member) of the board has the requisite familiarity with the subject of a particular manuscript. The editor must then obtain one or both reviews from non–board members, often called ad hoc reviewers or editorial consultants. (Also, some journals depend entirely on ad hoc reviewers.) Sometimes, the editor must make many inquiries before appropriate reviewers for a given manuscript are identified.

How do journals choose ad hoc reviewers? Often, the editors or editorial board members know of suitable candidates. Some journals keep databases of researchers who have served as reviewers or could do so; as well as noting areas of expertise, such databases sometimes include information on promptness and quality of reviews received. Editors often invite authors of works cited in the manuscript to serve as reviewers. They also search the literature on the topic to identify appropriate candidates. As discussed in the previous chapter, some journals allow authors to suggest potential reviewers—and let them list people they consider unsuited to serve as peer reviewers, for example because of conflicts of interest. (Editors get suspicious, though, when authors include in the latter list most of the researchers in their fields!) Also, when researchers who are invited to review a paper are not available, they typically are asked to identify others who are qualified to do so. And if you have a paper accepted by a journal, you may be added to its pool of potential reviewers.

Does the peer review system work? According to Bishop (1984, p. 45), "The answer to this question is a resounding, Yes! All editors, and most authors, will affirm that there is hardly a paper published that has not been improved, often substantially, by the revisions suggested by referees."

Most journals use anonymous reviewers. A few journals make the authors anonymous by deleting their names from the copies of manuscripts sent to reviewers. In general, experience seems to be in accord with that of the distinguished Canadian scientist J. A. Morrison, who said (1980): "It is occasionally argued that, to ensure fairness, authors should also be anonymous, even though that would be very difficult to arrange. Actually, editors encounter very few instances of unfairness and blatant bias expressed by referees; perhaps for 0.1 per cent or less of the manuscripts handled, an editor is obliged to discount the referee's comments."

If the reviewers have been chosen wisely, the reviews will be meaningful and the editor will be in a good position to arrive at a decision regarding publication of the manuscript. Also, whether the paper is accepted or not, the author will receive from the reviewers suggestions that can improve it. When the reviewers have returned the manuscripts, with their comments, the editor must face the moment of truth.

Peer review has been a subject of considerable research and reflection, and a number of international congresses have focused on the topic. Resources for those interested in peer review include books (Lock 1985; Godlee and Jefferson 2003) containing extensive bibliographies on the subject and the website of the International Congress on Peer Review and Biomedical Publication (www.peerreviewcongress.org/index.html).

THE EDITOR'S DECISION

Sometimes, the editor's decision is easy. If all reviewers advise "accept" with no or only slight revision, and all state solid reasons for their recommendations, the editor has no problem. Unfortunately, there are many instances in which the opinions of the reviewers are contradictory or unaccompanied by strong evidence. In such cases, the editor must either make the final decision or send the manuscript to one or more additional reviewers to determine whether a consensus can be established. The editor is likely to take the first approach if he or she is reasonably expert in the subject area of the manuscript and can thus serve as an additional reviewer; he or she is especially likely to do this if the detailed commentary of one reviewer is considerably more persuasive than that of the other. The second approach is obviously time-consuming and is used commonly by weak editors; however, any editor must use this approach if the manuscript concerns a subject with which he or she is not familiar. At journals with many more submissions than they can publish, even papers receiving all "accepts" may be rejected if strong arguments cannot be mustered for their inclusion (much as when a grant application is "approved but not funded").

The review process being completed, and the editor having made a decision, the author is now notified of the editor's decision. And it *is* the editor's decision. Editorial board members and ad hoc reviewers can only recommend; the final decision is and must be the editor's. This is especially true for those journals (the majority) that use anonymous reviewers. The decisions will be presented to the authors as though they were the editor's own, and indeed they are.

The editor's decision will be one of three general types, commonly expressed in one word as accept, reject, or modify. Commonly, one of these three decisions will be reached within 4 to 6 weeks after submission of the manuscript. If you are not advised of the editor's decision within 8 weeks, or provided with an explanation for the delay, do not be afraid to contact the journal. You have the right to expect a decision, or at least a report, within a reasonable length of time; also, your inquiry might bring to light a problem. Perhaps the editor's decision was made but notification did not reach you. If the delay was caused within the editor's office (usually by lack of response from one of the reviewers), your inquiry is likely to trigger an effort to resolve the problem, whatever it is.

Besides which, you should never be afraid to contact editors. With rare exceptions, editors are very nice people. Never consider them adversaries. They are on *your* side. Their only goal is to publish good science in understandable language. If that is not your goal also, you will indeed be dealing with a deadly adversary; however, if you share the same goal, you will find the editor to be a resolute ally. You are likely to receive advice and guidance that you could not possibly buy.

(© Aries Systems Corporation)

THE ACCEPT LETTER

Finally, you get the word. Suppose that the editor's letter announces that your manuscript has been accepted for publication. When you receive such a letter, you have every right to treat yourself to a glass of champagne or a hot fudge sundae or whatever you choose when you have cause both to celebrate and to admire yourself. The reason that such a celebration is appropriate is the relative rarity of the event. In the good journals (in biology at least), only about 5 percent of the papers are accepted as submitted.

THE MODIFY LETTER

More likely, you will receive from the editor a cover letter and two or more lists labeled "reviewers' comments." The letter may say something like, "Your manuscript has been reviewed, and it is being returned to you with the attached comments and suggestions. We believe these comments will help you improve your manuscript." This is the beginning phraseology of a typical modify letter. The letter may go on to say that the paper will be published if modified as requested, or it may say only that it will be reconsidered if the modifications are made.

By no means should you feel disconsolate when you receive such a letter. Realistically, you should not expect that rarest of all species, the accept letter without a request for modification. The vast majority of submitting authors will receive either a modify letter or a reject letter, so you should be pleased to receive the former rather than the latter.

When you receive a modify letter, examine it and the accompanying reviewers' comments carefully. (In many cases, the modify letter is a form letter, and it is the accompanying comments that are significant. Sometimes, however, the editor's letter contains specific guidance, such as regarding a point about which the reviewers disagree.) The big question now is whether you can, and are willing to, make the changes requested.

If all referees point to the same problem in a manuscript, almost certainly it *is* a problem. Occasionally, a referee may be biased, but hardly two or more simultaneously. If referees misunderstand, readers will as well. Thus, our advice is: If referees misunderstand the manuscript, find out what is wrong and correct it before resubmitting the manuscript to the same journal or another journal.

If the requested changes are relatively few and slight, you should go ahead and make them. As King Arthur used to say, "Don't get on your high horse unless you have a deep moat to cross."

If major revision is requested, however, you should step back and take a total look at your position. One of several circumstances is likely to exist.

First, the reviewers are right, and you now see that there are fundamental flaws in your paper. In that event, you should follow their directions and rewrite the manuscript accordingly.

Second, perhaps the reviewers have caught you off base on a point or two, but some of the criticism is invalid. In that event, you should rewrite the manuscript with two objectives in mind: Incorporate all of the suggested changes that you can reasonably accept, and try to beef up or clarify those points to which the reviewers (wrongly, in your opinion) took exception. Finally, and importantly, when you resubmit the revised manuscript, provide a letter indicating point by point what you did about the reviewers' comments.

Third, it is entirely possible that at least one reviewer and the editor seriously misread or misunderstood your manuscript, and you believe that their criticisms are almost totally erroneous. In that event, you have two alternatives. The first, and more feasible, is to submit the manuscript to another journal, hoping that your manuscript will be judged more suitably. If, however, you have strong reasons for wanting to publish that particular manuscript in that particular journal, do not back off; resubmit the manuscript. In this case, however, you should use all the tact at your command. Not only must you give a point-by-point

"Thank you for your detailed and lengthy criticism of my manuscript. I will be sure to incorporate your suggestions in my next draft."

(© Vivian S. Hixson, reproduced by permission)

rebuttal of the reviewers' comments; you must do it in a way that is not antagonistic. Remember that the editor is trying hard, probably without pay, to reach a *scientific* decision. If you start your covering letter by saying that the reviewers, whom the editor obviously has selected, are "stupid" (yes, such letters exist), we will give you 100 to 1 that your manuscript will be immediately returned without further consideration. On the other hand, *every* editor knows that *every* reviewer can be wrong and in time (Murphy's Law) will be wrong. Therefore, if you calmly point out to the editor exactly why you are right and the reviewer is wrong (*never* say the editor is wrong), the editor is likely to accept your manuscript at that point or, at least, send it to one or more additional reviewers for further consideration.

If you do decide to revise and resubmit the manuscript, try very hard to meet whatever deadline the editor establishes. Most editors do set deadlines. Obviously, many manuscripts returned for revision are not resubmitted to the same journal; hence, the journal's records can be cleared of deadwood by considering manuscripts to be withdrawn after the deadline passes.

If you meet the editor's deadline, he or she may accept the manuscript forthwith. Or, if the modification has been substantial, the editor may return it to the same reviewers. If you have met, or defended your paper against, the previous criticism, your manuscript probably will be accepted.

On the other hand, if you fail to meet the deadline, your revised manuscript may be treated as a new manuscript and again be subjected to full review, possibly by a different set of reviewers. It is wise to avoid this double jeopardy, plus additional review time, by carefully observing the editor's deadline if it is at all possible to do so. If you believe that you cannot meet the deadline, immediately explain the situation to the editor; the deadline might then be extended.

When you submit a revised manuscript, make it easy for the editor to identify the changes. For example, if the editor supplied a numbered list of revisions to make, state, by number, how each was addressed. Perhaps use the Track Changes feature of Word to show your revisions. Or if the editor asked you to indicate your revisions in another way, carefully follow the instructions. Clearly identifying the changes made can speed the final decision about your paper. It also can help earn you a reputation as a good author to work with—a fact that can facilitate further interactions with the editorial office.

THE REJECT LETTER

Now let us suppose that you get a reject letter. (Almost all editors say "unacceptable" or "unacceptable in its present form"; seldom is the harsh word "reject" used.) Before you begin to weep, do two things. First, remind yourself that you have a lot of company; most of the good journals have rejection rates of 50 percent or more. Second, read the reject letter *carefully* because, like modify letters, there are different types of rejection.

Many editors would class rejections in one of four ways. First, there is (rarely) the total rejection, the type of manuscript that the editor "never wants to see again" (a phrase that one undiplomatic editor put into a reject letter). Second, and much more common, there is the type of manuscript that contains some

(PEANUTS © 1987 Peanuts Worldwide LLC. Dist. By UNIVERSAL UCLICK. Reprinted with permission. All rights reserved.)

useful data but in which the data are seriously flawed. The editor probably would reconsider such a manuscript if it were considerably revised and resubmitted, but the editor does not recommend resubmission. Third, there is the type of manuscript that is basically acceptable, except for a defect in the experimental work—the lack of a control experiment, perhaps—or except for a major defect in the *manuscript* (the data being acceptable). Fourth, in the case of highly competitive journals, there is the manuscript presenting research that, although sound, is not deemed important enough or of broad enough interest for inclusion.

If your "rejection" is of the third type, you might well do the necessary repairs, as described in the reviewers' comments, and resubmit a revised version to the same journal. If you can add that control experiment, as requested by the editor, the new version might be accepted. (Many editors reject a paper that requires additional experimentation, even though it might be easy to modify the paper to acceptability.) Or, if you make the requested major change in the manuscript (for example, totally rewriting the discussion or converting a full paper to a note), your resubmitted manuscript is quite likely to be accepted.

If your rejection is of the second type (because of seriously flawed data, according to the editor's reject letter and the reviewers' comments), you should probably not resubmit the same manuscript to the same journal, *unless* you can make a convincing case to the editor that the reviewers seriously misjudged your manuscript. You might, however, keep the manuscript until it can be buttressed with more extensive evidence and more clear-cut conclusions. Resubmission of such a "new" manuscript to the same journal would then be a reasonable option. Your cover letter should mention the previous manuscript and should state briefly the nature of the new material.

If your rejection is of the first (total) or fourth (priority-based) type, it would be pointless to resubmit the manuscript to the same journal or even to argue about it. If the manuscript is really bad, you probably should not (re)submit it anywhere, for fear that publication might damage your reputation. If there is work in it that can be salvaged, incorporate those portions into a new manuscript and try again, but in a different journal. If the work was deemed competent but not of high enough priority, take advantage of any useful suggestions from the reviewers, and promptly submit the manuscript to another journal. Your manuscript may well find ready acceptance in a more specialized or otherwise less competitive venue.

Cheer up. You may someday have enough rejection letters to paper a wall with them. You may even begin to appreciate the delicate phrasing that is sometimes used. Could a letter such as the following possibly hurt? (This is reputedly a rejection slip from a Chinese economics journal.)

> We have read your manuscript with boundless delight. If we were to publish your paper, it would be impossible for us to publish any work of a lower standard. As it is unthinkable that, in the next thousand years, we shall see its equal, we are, to our regret, compelled to return your divine composition, and to beg you a thousand times to overlook our short sight and timidity.

EDITORS AS GATEKEEPERS

Perhaps the most important point to remember, whether dealing with a modify decision or a reject one, is that the editor is a mediator between you and the reviewers. If you deal with editors respectfully, and if you can defend your work scientifically, most of your "modifies" and even your "rejects" will in time become published papers. The editor and the reviewers are usually on your side. Their primary function is to help you express yourself effectively and provide you with an assessment of the science involved. It is to your advantage to cooperate with them in all ways possible. The possible outcomes of the editorial process were neatly described by Morgan (1986): "The modern metaphor for editing would be a car wash through which all cars headed for a goal must pass. Very dirty cars are turned away; dirty cars emerge much cleaner, while clean cars are little changed."

Were it not for the gatekeeper role so valiantly maintained by editors, our scientific journals would soon be reduced to unintelligible gibberish.

No matter how you are treated by editors, try somehow to maintain a bit of sympathy for that benighted profession. H. L. Mencken wrote a letter dated January 25, 1936, to William Saroyan, saying, "I note what you say about your aspiration to edit a magazine. I am sending you by this mail a six-chambered revolver. Load it and fire every one into your head. You will thank me after you get to Hell and learn from other editors how dreadful their job was on earth."

CHAPTER 22

The Publishing Process (How to Deal with Proofs)—and After Publication

Proofread carefully to see if you any words out.

—Anonymous

THE COPYEDITING AND PROOFING PROCESSES

The following is a brief description of the process that your manuscript follows after it has been accepted for publication.

The manuscript usually goes through a copyediting procedure during which errors in spelling and grammar are corrected. In addition, the copy editor will standardize all abbreviations, units of measure, punctuation, and spelling in accord with the style of the journal in which your manuscript is to be published. At some journals, copy editors also revise writing to increase readability, for example by improving sentence structure and making wording more concise. Many English-language journals with sufficient staff to do so devote extra effort to copyediting papers by non-native speakers of English, in order to promote clear international communication. The copy editor may direct questions to you if any of your wording is not clear or if any additional information is needed. These questions may appear as *author queries* written on or accompanying the *proofs* (copies of typeset material) sent to the author. Alternatively, the queries may appear on or with the copyedited manuscript, if the journal sends it to the author for approval before preparing the proof.

Typically, the edited version of the electronic file that you provided is loaded into a computer system that can communicate with a typesetting system, which will produce the proofs of your article. The copy editor or compositor keyboards codes that indicate the typefaces and page layout.

The output of this effort is your set of proofs, which is then returned to you so that you may check the editorial work that has been done on your article, check for typographical errors, and answer any questions by the copy editor. Commonly, you will receive the proof of your article as a PDF file.

Finally, someone at the journal will keyboard the corrections that you make on your proofs. The final version will become the type that you see on the pages of the journal.

WHY PROOFS ARE SENT TO AUTHORS

Some authors seem to forget their manuscripts as soon as they are accepted for publication, paying little attention to the proofs when they arrive and assuming that their papers will magically appear in the journals, without error.

Why are proofs sent to authors? Authors are provided with proofs of their papers for one main reason: to check the accuracy of the type composition. In other words, you should examine the proofs carefully for typographical errors. Even if you carefully proofread and spell-checked your paper before submitting it, errors can remain or can occur when editorial changes are input. And sometimes the typesetting process mysteriously converts Greek letters into squiggles, cuts off lines of text, or causes other mischief. No matter how perfect your manuscript might be, it is only the printed version in the journal that counts. If the printed article contains serious errors, all kinds of later problems can develop, not the least of which may be serious damage to your reputation.

The damage can be real in that many errors can destroy comprehension. Something as minor as a misplaced decimal point can sometimes make a published paper almost useless. In this world, we can be sure of only three things: death, taxes, and typographical errors.

MISSPELLED WORDS

Even if the error does not greatly affect comprehension, it won't do your reputation much good if it turns out to be funny. Readers will know what you mean if your paper refers to a "nosocomical infection," and they will get a laugh out of it, but *you* won't think it is funny.

A major laboratory-supply corporation submitted an ad with a huge boldface headline proclaiming that "Quality is consistant because we care." We certainly hope they cared more about the quality of their products than they did about the quality of their spelling.

Although all of us in publishing occasionally lose sleep worrying about typographical errors, we can take comfort in the realization that whatever slips by our eye is probably less grievous than some of the monumental errors committed by our publishing predecessors.

An all-time favorite error occurred in a Bible published in England in 1631. The Seventh Commandment read: "Thou shalt commit adultery." We understand that Christianity became very popular indeed after publication of that edition. If that statement seems blasphemous, we need only refer you to another edition of the Bible, printed in 1653, in which appears the line: "Know ye that the unrighteous shall inherit the kingdom of God."

If you read proofs in the same way and at the same speed that you ordinarily read scientific papers, you will probably miss 90 percent of the typographical errors.

The best way to read proofs is, first, *read* them and, second, *study* them. The reading will miss 90 percent of the errors, but it will catch errors of *omission.* If the printer has dropped a line, reading for comprehension is the only likely way to catch it. Alternatively, or *additionally,* it can be helpful for two people to read the proof, one reading aloud while the other follows the manuscript.

To catch most errors, however, you must slowly examine each word. If you let your eye jump from one group of words to the next, as it does in normal reading, you will not catch very many misspellings. Especially, you should study the technical terms. A good keyboarder might be able to type the word "cherry" 100 times without error; however, there was a proof in which the word "*Escherichia*" was misspelled 21 consecutive times (in four different ways). One might also wonder about the possible uses for a chemical whose formula was printed as $C_{12}H_6Q_3$. One way to look at each word without distraction is to read the proof backward, from last word to first.

As a safeguard, consider having someone else review the proof, in addition to doing so yourself. But do not delegate the proofreading solely to others, lest you suffer the plight of a colleague of ours who, tired of the publication process, had an office worker review the proof. Only after the journal was published did the colleague find that the article title contained a misspelling.

We mentioned the havoc that could occur from a misplaced decimal point. This observation leads to a general rule in proofreading. Examine each and every number carefully. Be especially careful in proofing the tables. This rule is important for two reasons. First, errors frequently occur in keyboarding numbers, especially in tabular material. Second, you are the *only* person who can catch such errors. Most spelling errors are caught in the printer's proof room or in the journal's editorial office. However, professional proofreaders catch errors by "eyeballing" the proofs; the proofreader has no way of knowing that a "16" should be a "61."

MARKING THE CORRECTIONS

Like much else in scientific publishing, correction of proofs has been changing in the electronic era. No longer do authors receive galley proofs (long strips of type) to correct before page proofs are prepared. And rather than being sent proofs by mail, authors receive them electronically or access them through websites. Accordingly, procedures for indicating corrections have been evolving. Be sure to follow the current instructions that the journal provides with the proof.

The long-established procedure is to mark each error *twice* on a hard copy of a proof, once at the point where it occurs and once in the margin opposite where it occurs. The compositor uses the margin marks to find the errors, as a correction indicated only in the body of the text can easily go unnoticed. Standard proofreading marks, the most common of which are listed in Table 22.1, should be used to indicate corrections. Normally, if you are to print out a hard copy and indicate corrections on it, the publisher provides a list of such marks along with the proof. Learning the main such marks can facilitate reviewing the proofs of your papers and aids in proofing typeset versions of other items you write.

Other options have been developing. Some journals, for example, ask authors to use tools in Adobe Acrobat to indicate corrections on PDF versions of the proof. Some have their own online proofreading systems for authors to use.

Whatever approach is used, return the proofs quickly, by the deadline from the journal. Failure to do so disrupts the publication schedule of the journal and can result in delay or even withdrawal of your paper. If you think you might be unreachable when the proofs become available, inform the journal, so the timetable can be revised or the proofs can be sent to a coauthor or other colleague to review.

ADDITIONS TO THE PROOFS

Early in this chapter, we stated that authors are sent proofs so that they can check the accuracy of the typesetting. Stated negatively, the proof stage is *not* the time for revision, rewriting, rephrasing, addition of more recent material, or any other significant change from the final edited manuscript. There are three good reasons why you should not make substantial changes in the proofs.

First, an ethical consideration: Since neither proofs nor changes in the proofs are seen by the editor unless the journal is a small one-person operation, it simply is not proper to make substantive changes. The paper approved by the editor, after peer review, is the one that should be printed, not some new version containing material not seen by the editor and the reviewers.

Table 22.1. Frequently used proofreaders' marks

Instruction	Mark in text	Mark in margin
Capitalize	Hela cells	cap
Make lower case	the Penicillin reaction	l.c.
Delete	a very good reaction	ℓ
Close up	Mac Donald reaction	⁀‿
Insert space	lymphnode cells	#
Start new paragraph	in the cells. The next	¶
Insert comma	in the cells after which	^,
Insert semicolon	in the cells however	^;
Insert hyphen	well known event	=
Insert period	in the cells Then	⊙
Insert word	in cells	#the#
Transpose	proofraeder	tr
Subscript	CO2	2
Superscript	32P	32
Set in roman type	The *bacterium* was	rom
Set in italic type	P. aeruginosa cells	ital
Set in boldface type	Results	b.f.
Let it stand	a very good reaction	stet

Second, it is not wise to disturb typeset material, unless it is really necessary, because new typographical errors may be introduced.

Third, corrections can be expensive. Therefore you should not abuse the publisher (possibly a scientific society of which you are an otherwise loyal member) by requesting unessential changes; in addition, you just might receive a substantial bill for author's alterations. Most journals absorb the cost of a reasonable number of author's alterations, but many, especially those with managing editors or business managers, will sooner or later crack down on you if you are patently guilty of excessive alteration of the proofs.

One type of addition to the proofs is frequently allowed. The need arises when a paper on the same or a related subject is published while yours is in process. In light of the new study, you might be tempted to rewrite several portions of your paper. You must resist this temptation, for the reasons stated previously. What you should do is prepare a short addendum in proof (at most a few sentences), describing the general nature of the new work and giving the literature reference. If the editor approves including it, the addendum can then be printed at the end without disturbing the body of the paper.

ADDITION OF REFERENCES

Quite commonly, a new paper appears that you would like to add to your references; in doing so you would not need to make any appreciable change in the text, other than adding a few words, perhaps, and the number of the new reference. If you are unsure how the journal would like you to proceed, consult its editorial office.

If the journal employs the numbered, alphabetized reference system, you may be asked to add the new reference with an "a" number. For example, if the new reference would alphabetically fall between references 16 and 17, the new reference would be listed as "16a." In that way, the numbering of the rest of the list need not be changed. Thus would be avoided the cost and the potential for error of renumbering the references in the reference list and the text. An analogous procedure may be requested for references in the citation order system. Conveniently, if the new reference is cited in an addendum, it would appear last in the citation order system, thus not disrupting the rest of the reference list.

PROOFING THE ILLUSTRATIONS

It is important to examine carefully the proofs of the illustrations. In the era when authors submitted photographic prints and other hard-copy illustrations, one needed to check carefully for quality of reproduction. Now that illustrations are being submitted electronically, such need has diminished. Nevertheless, checking illustrations remains important. Make sure that all illustrations are present, complete, and appropriately placed. Also make sure that electronic gremlins have not somehow disturbed or distorted the illustrations. If you perceive problems with the illustrations, report them as instructed by the journal.

WHEN TO COMPLAIN

If you have learned nothing else from this chapter, we trust that you now know that *you* must provide quality control. Too many authors have complained after the fact (after publication) without ever realizing that only they could have prevented whatever it is they are complaining about. For example, authors many times have complained that their pictures have been printed upside down or sideways. When such complaints have been checked, it has commonly been found that the author failed to note that the photo was oriented incorrectly in the proof.

So, if you are going to complain, do it at the proof stage. And, believe it or not, your complaint is likely to be received graciously. Publishers have invested heavily in setting the specifications that can provide quality reproduction. They need your quality control, however, to ensure that their money is not wasted.

Good journals are printed by good printers, hired by good publishers. The published paper will have your name on it, but the reputations of both the publisher and the printer are also at stake. They expect you to work with them in producing a superior product. Likewise, if a journal is solely electronic, the publisher wants to ensure that the product is of high quality and depends on your collaboration in that regard.

Because managing editors of such journals must protect the integrity of the product, those we have known would *never* hire a printer exclusively on the basis of low bids. John Ruskin was no doubt right when he said, "There is hardly anything in the world that somebody cannot make a little worse and sell a little cheaper, and the people who consider price only are this person's lawful prey."

A sign in a printing shop made the same point:

PRICE
QUALITY
SERVICE

Pick any two of the above

REPRINTS

Customarily, authors have received with their proofs a form for ordering hard-copy reprints of their articles. Older scientists remember the days—before widespread electronic access to journal articles, and indeed before widespread access to photocopying—when obtaining reprints from authors served as an important way to keep up with the literature. As well as giving researchers access to articles, reprint requests helped authors learn who was interested in their work (Wiley 2009).

Today, reprints are much less a part of scientific culture. Nevertheless, they sometimes remain worth ordering (for example, to share with colleagues in countries with limited access to the journal literature, to have available at poster presentations or job interviews, or to impress your mother or prospective spouse). Norms regarding reprints differ among research fields. If you are unfamiliar with those in yours, seek colleagues' guidance on whether to order reprints and, if so, how to use them.

Some journals also make available "electronic reprints," which allow authors to grant one-time electronic access to their articles. Someday scientists may ask why articles that authors share are called reprints at all.

PUBLICIZING AND ARCHIVING YOUR PAPER

In the era when paper reprints prevailed, scientists commonly sent them to colleagues worldwide soon after publishing a paper. Today many scientists alert others to their new articles largely through social media. Depending on their preferences and the scientific cultures where they work, they may, for instance, tweet news of the publication on Twitter and post it on Facebook. They also may add listings (and links) to their ORCID records, LinkedIn profiles, and profiles on science-related networking sites such as ResearchGate. When you publish an article, such steps—and the follow-up by those who thus notice the article and inform others in their networks—can inform those potentially interested in your paper.

General media too can aid in publicizing your newly published research, both to the public and to fellow researchers. Many universities and other research institutions have public information officers (PIOs) whose role is largely to publicize the research done there. When you have a paper accepted, alert a PIO at your institution. He or she can then consider whether to publicize the research to journalists and others, for instance through news releases, institutional websites and publications, and use of social media networks. PIOs know that some journals place articles under *embargoes*; in other words, research reported in them is not to appear in the media until the release date for the issue in which they are published. A PIO can aid in obtaining timely coverage without violating embargoes. Advice on working with PIOs appears in an article by Tracy Vence (2015) in *The Scientist*.

Publicizing a newly published paper—through social media, mass media, or other means—can benefit a scientist in multiple ways, notes PIO Matt Shipman (2015). For example, it can lead to citations, please funding agencies, engender collaborations, and more generally expand one's professional network. It also may attract potential graduate students—or perhaps, earlier in one's career, attract attention of employers or postdoctoral-fellowship sites. And if

"I haven't been in any academic journals but I do get my Tweets re-Tweeted a lot."

(www.CartoonStock.com)

your findings have applications outside the research sector, publicity may bring them to the attention of those who can use them there. Indeed, if research is supported with public funds, scientists may be morally obligated to get the word out. In fact, some public and other sources of research funds require that grant recipients make their work openly available.

Whether required or not, making your journal articles (and reports based on them) widely available can benefit science, society, and your career. Follow, of course, the policies of relevant funding agencies regarding public access to papers resulting from your research. (An example is the U.S. National Institutes of Health public access policy, publicaccess.nih.gov/policy.htm.) If your university or other institution has an institutional archive, explore depositing electronic copies of your publication there. Also consider linking publications to your own website or your curriculum vitae. The SHERPA/RoMEO website, www.sherpa.ac.uk/romeo, includes information on journal publishers' policies on self-archiving.

In short, publishing a journal article, though a major accomplishment, is not the last step in getting word out about your research. In ways, it is just the beginning. You have invested great effort in doing the research, writing a paper about it, and publishing the paper. Now take the additional steps to help ensure that, in the broadest sense, your paper has maximum impact.

CELEBRATING PUBLICATION

As noted, publishing a scientific paper is a major accomplishment. By the time a paper comes out, you may well be working on the next paper—or even the one after. But take some time to celebrate your success. Some scientists or research groups have traditions for doing so. For example, some gather for a celebration dinner and post a photo taken there. Some treat themselves to a good bottle of wine and collect the labels as reminders of their success. Some may attend a concert, play, or athletic event and keep the tickets or program as mementos. Perhaps best of all is to spend some time—perhaps at a favorite park—with the family members or friends who tolerated our absences, insecurities, and complaints as we struggled to write about and publish the work. Whatever you choose, you deserve to celebrate. Congratulations on publishing your paper!

PART V

Doing Other Writing for Publication

CHAPTER 23

How to Write a Review Paper

Review articles constitute a form of original research, albeit done in the library rather than in the laboratory or at the bedside.

—Bruce P. Squires

CHARACTERISTICS OF A REVIEW PAPER

A review paper is *not* an original publication in the usual sense, though it can be valuable scholarship. On occasion, a review will contain new data (from the author's own laboratory) that have not yet appeared in a primary journal. However, the purpose of a review paper is to review previously published literature and to put it into perspective.

A review paper is oftentimes long, often ranging between 10 and 50 published pages. (Some journals now print short "mini reviews.") The subject is fairly general, compared to that of research papers, and the literature review is, of course, the principal product. However, the really good review papers are much more than annotated bibliographies. They offer critical evaluation of the published literature and often provide important conclusions based on that literature.

The organization of a review paper usually differs from that of a research paper. The introduction, materials and methods, results, and discussion arrangement traditionally has not been used for the review paper. However, some review papers are prepared more or less in the IMRAD format; for example, they may contain a methods section describing how the literature review was done.

If you have previously written research papers and are now about to write your first review paper, it might help you conceptually if you visualize the review paper as a research paper, as follows: Greatly expand the introduction;

delete the materials and methods (unless original data are being presented or you will say how you identified and chose the literature to include); delete the results; and expand the discussion.

Actually, you may have already written the equivalent of many review papers. In format, a review paper is not very different from a well-organized term paper or literature review section of a thesis.

As in a research paper, however, it is the *organization* of the review paper that is important. The writing will almost take care of itself if you can get the thing organized.

PREPARING AN OUTLINE

Unlike for research papers, there is no prescribed organization for conventional review papers. Therefore, you will have to develop your own. A cardinal rule for writing a review paper is *prepare an outline*.

The outline must be prepared carefully. It will assist you in organizing your paper, which is all-important. If your review is organized properly, the overall scope of the review will be well defined and the integral parts will fit together in logical order.

Obviously, you must prepare the outline *before* you start writing. Moreover, *before* you start writing, it is wise to determine whether a journal (either a review journal or a primary journal that includes review articles) would be interested in considering a review article that you submit on the topic. Possibly, the editor will want to limit or expand the scope of your proposed review or add or delete specific subtopics. Or perhaps the journal is already publishing a review on the subject, in which case you should direct your effort elsewhere.

Not only is the outline essential for the preparer of the review, it is also very useful to potential readers of the review. Therefore, many review journals print the outline at the beginning of the article, where it serves as a convenient table of contents for prospective readers.

Also to guide readers, review papers make considerable use of subheadings (which, if an outline is published, correspond to the subjects it lists). For example, the review paper "Mechanics of Cytokinesis in Eukaryotes" by Thomas D. Pollard (*Curr. Opin. Cell. Biol.* 22:50–56, 2010) contains the following subheadings:

Introduction
Origins of cytokinesis genes
Mechanisms specifying the position of the division plane
    Fission yeast
    Budding yeast
    Animal cells

Mechanism of contractile ring assembly
  Fission yeast
  Animal cells
Architecture of the ring
Mechanism of constriction and disassembly of the contractile ring
  Actin filaments
  Myosin-II
  Mechanism of constriction
  Sources of drag
  Modeling
Conclusions

In 2015, Pollard, the author of this review paper, received the National Academy of Sciences Award for Scientific Reviewing "for his many review articles describing the molecular mechanisms of the protein actin in cell motility and cell division"; it was noted that these articles "have been cited hundreds and even thousands of times." This award, given in different years to authors in different fields, has been presented since 1979. Information on recipients appears at www.nasonline.org/programs/awards/scientific-reviewing.html, on the (U.S.) National Academy of Sciences website. To see some review papers by masters, look on this site to identify recipients in your field, and then search the literature to find their reviews.

TYPES OF REVIEWS

Before actually writing a review, you also need to determine the requirements of the journal to which you plan to submit the manuscript. Some journals demand critical evaluation of the literature, whereas others are more concerned with bibliographic completeness. There are also matters of organization, style, and emphasis that you should consider before you proceed very far.

By and large, the old-line review journals prefer, and some demand, authoritative and critical evaluations of the published literature on a subject. Many of the "book" series ("Annual Review of," "Recent Advances in," "Yearbook of," etc.), however, publish reviews designed to compile and to annotate but not necessarily to evaluate the papers published on a particular subject during a defined time period. Some active areas of research are reviewed yearly. Both of these types of review papers serve a purpose, but the different purposes need to be recognized.

At one time, review papers tended to present historical analyses. In fact, the reviews were often organized chronologically. Although this type of review is now less common, one should not deduce that the history of science has become less important. There is still a place for history.

Today, however, most review media prefer either "state of the art" reviews or reviews that provide a new understanding of a rapidly moving field. Mainly the recent literature on the subject is catalogued or evaluated. If you are reviewing a subject that has not previously been reviewed or one in which misunderstandings or polemics have developed, a bit more coverage of the historical foundations would be appropriate. If the subject has been effectively reviewed before, the starting point for your review might well be the date of the previous review (not publication date, but the date up to which the literature has been reviewed). And, of course, your review should begin by citing the previous review.

Another type of review paper, known as a *systematic review article,* has become common in some fields. A systematic review addresses "a clearly formulated question" and "uses systematic and explicit methods to identify, select, and critically appraise relevant research, and to collect and analyse data from the studies that are included in the review" (community.cochrane.org/glossary/5#letters). Commonly, systematic review articles follow a variant of the IMRAD format; for example, they include a methods section specifying such items as databases searched, search terms used, dates and languages included, and criteria for including and excluding studies. Academic librarians, some of whom specialize in literature searching for systematic reviews, can be well worth consulting when preparing such a review. Written sources of guidance for preparing systematic review articles include the PRISMA statement (Moher et al. 2009).

WRITING FOR THE AUDIENCE

Another basic difference between review papers and primary papers is the *audience.* The primary paper is highly specialized, and so is its audience (mainly peers of the author). The review paper will probably cover a number of highly specialized subjects in your field, and so the review will be read by many peers. The review paper will also be read by many people in related fields, because the reading of good review papers is the best way to keep up in one's broad areas of interest—or to start preparing to enter related areas of research. Finally, review papers are valuable in teaching, so student use is likely to be high.

Because the review paper is likely to have a wide and varied audience, your style of writing should be much less technical than for a research paper. Jargon and specialized abbreviations must be eliminated or carefully explained. Your writing style should be expansive rather than telegraphic (condensed).

IMPORTANCE OF INTRODUCTORY PARAGRAPHS

Readers are much influenced by the introduction of a review paper. They are likely to decide whether to read further on the basis of what they find in the first few paragraphs (if they haven't already been repelled by the title).

Readers are also influenced by the first paragraph of each major section of a review, deciding whether to read, skim, or skip the rest of the section depending on what they find in the first paragraph. If first paragraphs are well written, all readers, including the skimmers and skippers, will be able to achieve some comprehension of the subject.

IMPORTANCE OF CONCLUSIONS

Because the review paper typically covers a wide subject for a wide audience, a form of conclusions is worth taking the trouble to write. Doing so is especially important for a highly technical, advanced, or obscure subject. Painful compromises must sometimes be made if one really tries to summarize a difficult subject to the satisfaction of both expert and amateur. Yet, good summaries and simplifications will in time find their way into textbooks and mean a great deal to students yet to come.

CHAPTER 24

How to Write Opinion (Letters to the Editor, Editorials, and Book Reviews)

It is hard enough to remember my opinions, without also remembering my reasons for them!

—Friedrich Nietzsche

WRITING INFORMED OPINION

As you become known in your field, editors of journals and other publications may invite you to write pieces expressing your professional judgment. In particular, you may be asked to write editorials and book reviews. Chances to write the latter also may arise earlier in your career. And whatever your seniority (or lack thereof), you may submit letters to the editor for potential publication or posting.

All these pieces express opinion. But not just any opinion: your scientifically informed opinion. Although sometimes allowing more creativity in writing style, they should display the same rigor as a scientific paper. Evidence should support views, and logic should be tight. In short, scientific opinion pieces should clearly evidence the mind of a researcher.

LETTERS TO THE EDITOR

Many journals print or post letters to the editor. Usually, not all letters received are published.

Often, letters comment on papers recently published in the journal, thus serving as post-publication peer review. Sometimes, they deal independently with issues of professional interest to readers. In some journals, brief research reports or case reports can appear as letters; an editor who decides not to publish a paper may offer to publish a condensed version as a letter to the editor. When a letter comments on a paper, the authors of the paper may have the opportunity to prepare a reply for publication.

Before drafting a letter, check the journal's instructions, which commonly appear in the letters section of the journal and on the journal website. Among items that the instructions may specify are maximum length, number of figures and tables allowed, number of references allowed, and acceptable means of submission. Increasingly, journals have been requesting or requiring that letters be submitted electronically. Some journals' websites include a section through which letters can be submitted.

If you are writing a letter to the editor about a published article, submit it shortly after the article appeared. Some journals refuse to consider for publication those letters received after a stated interval. If you criticize an article, do so in a constructive and respectful tone. (Remember: The author might peer review your next scientific paper or grant proposal.) Similarly, if you are responding to a letter noting a possible shortcoming of your work, word your reply calmly—no matter what your initial reaction might have been.

Especially because of limitations in length, word your letter concisely, in keeping with principles presented later in this book in the section on scientific style. Focus on a single point (or a group of closely related points), and relate the other content to that central focus. Whatever your message, support it clearly. Your letter may then be a fine addition to the literature.

EDITORIALS

Some journals include invited editorials and other opinion pieces by scientists. In addition, scientists sometimes write opinion pieces for professional venues such as *The Scientist,* for op-ed pages of newspapers, or for other popular venues.

Invited editorials in journals can include both "perspective editorials" and "persuasive editorials." A perspective editorial provides context for and comments on a scientific paper in the same issue of the journal. Often, a scientist who peer reviewed the paper is invited to write it. The beginning of such an editorial commonly resembles a miniature review paper on the subject. The end can then serve somewhat like an independently written discussion section—noting, for example, strengths and limitations of the research reported in the paper and discussing implications. For a perspective editorial to appear in the same issue as the paper it comments on, it may need to be submitted quickly. Therefore, along with the honor of being invited to write such a piece you might receive a stringent deadline.

A persuasive editorial, in a journal or elsewhere, argues for a specific point of view, for example on science policy. How to structure your argument can depend on your audience. If your audience seems largely to agree with your main point, presenting it early and then supporting it can be most effective. If, however, many readers are likely to be opposed initially, you might gain greatest agreement by starting with mutually supported ideas and relatively unexceptionable data and then showing how they lead to your conclusions. Whatever your approach, include arguments for and against your point of view and competing points of view. Acknowledging other viewpoints and showing that yours is superior is scientifically sounder, and thus more credible, than acting as if other viewpoints do not exist.

Some journals publish unsolicited opinion pieces, sometimes called sounding boards. The principles of writing them tend to be much the same as for writing persuasive editorials. For guidelines on writing such items, consult the journal's instructions to authors. Similarly, if you wish to submit an opinion piece to a newspaper op-ed page or other popular venue, check the publication's requirements by looking at its website or contacting its editorial office.

BOOK (AND OTHER MEDIA) REVIEWS

Textbooks. Reference books. Specialized monographs for scientists. Trade books for the public. Science abounds with books. And many journals, magazines, and other publications include reviews of books on science. As well as helping readers choose books to obtain or consult, book reviews can inform readers by sharing content from the books. They also can provide useful feedback to authors and publishers and help guide future authors. Reviews of other media, such as journals and electronic resources, can serve similar functions. Regardless of whether a book or other item is reviewed, the principles are much the same. Thus, guidelines for writing book reviews apply in general to other reviews.

At journals, book review editors typically take the initiative in recruiting reviewers. However, they usually are glad to have potential reviewers volunteer, either to be approached as needed or to review specific books. Of course, if you have a conflict of interest (for example, because a book is by a close colleague), you should not offer to review the book or accept an invitation to do so.

A good review should both describe and evaluate the book. Among questions it may address are the following (Gastel 1991): What is the goal of the book, and how well does the book accomplish it? From what context did the book emerge? What is the background of the authors or editors? What is the scope of the book, and how is the content organized? What main points does the book make? If the book has special features, what are they? What are the strengths and weaknesses of the book? How does the book compare with other books on the same topic or with previous editions of the book? Who would find the book valuable?

Normally, answering these questions entails reading the book thoroughly. For a reference work, however, sampling the content is more feasible and better reflects the intended use. If you take such an approach, consider drawing on your skills in research design in determining how to proceed.

To facilitate writing, take notes as you read or mark passages of interest in the book. Write down ideas for points to make as they occur to you. To help formulate your ideas, perhaps tell someone about the book.

Although some journals feature structured book reviews, with standardized headings for specified types of content, the reviewer generally can choose how to organize the book review. One format that can work well is a variant of the IMRAD (introduction, methods, results, and discussion) structure commonly used for scientific papers. In this format, the "introduction" presents an opening comment on the book, the "results" describes the book, and the "discussion" evaluates it. No "methods" section may be needed if you read the book from cover to cover and did not otherwise test the book. But if, for example, you systematically sampled content in a reference book, you would summarize your procedure in the "methods."

A review is not an advertisement and should not gush with praise. Neither should it nitpick or ridicule. Rather, it should have a reasoned tone. By presenting information about the book and drawing careful conclusions, you will serve well the readers of your review.

CHAPTER 25

How to Write a Book Chapter or a Book

I'm writing a book. I've got the page numbers done.

—Steven Wright

HOW TO WRITE A BOOK CHAPTER

Congratulations! You have been invited to write a chapter in a multiauthored book. Here is one more sign that you have attained visibility in your field. Enjoy the compliment, and accept the invitation if you have the time to prepare the chapter well and submit it promptly. If you cannot write the chapter, recommend a peer if possible.

If you agree to write a chapter, be sure that the editor provides thorough instructions. Follow the instructions carefully; only if chapters are of the proper scope, length, and format, and only if they are submitted on time, can the book be published without undue difficulty and delay. If events arise that may slow submission of your chapter, tell the editor immediately so plans can be revised if needed.

In many cases, writing a book chapter is much like writing a review paper. If you are writing a chapter that summarizes knowledge on a topic, follow relevant advice from Chapter 23, "How to Write a Review Paper." In particular, plan the chapter carefully. Time invested in organizing the chapter can later save much time in writing.

After submitting the chapter, you may receive queries from the copy editor (for example, requests for clarification of points). You also may receive an edited manuscript and then page proofs to review. So as not to disrupt the production schedule, take care to respond by the deadline. If you will be unreachable for a substantial time while a chapter is in press, tell the editor so alternative plans can be made.

WHY (OR WHY NOT) TO WRITE A BOOK

There can be many good reasons to write a book. A monograph focusing on a specialized technical topic can aid fellow scientists. A handbook can assist scientists and those applying science. A textbook can greatly help students of science. A work of popular scientific nonfiction can interest and enlighten general readers, including those in fields of science other than your own.

There also can be good reasons not to write a book—or not to do so at present. In most fields of science, scientific papers (not books) are the currency of advancement. Thus, it can be unwise to spend time writing a book early in one's career. Of course, writing a book takes much effort and so should not be pursued without careful reflection first.

As for the monetary aspect: A widely used textbook or bestselling work of scientific nonfiction can net the author a nice sum. Most books in the sciences, however, earn the author relatively little—sometimes less than the author spent preparing the book. Thus, only if the psychological rewards would suffice should one embark on writing a book.

HOW TO FIND A PUBLISHER

Sometimes the publisher finds you. At companies publishing books in the sciences, editors keep track of science, for instance by attending scientific conferences. Thus, an editor may approach you about the possibility of writing a book.

If you are the one with the idea, see which publishers have published good books on topics related to yours. These publishers are most likely to accept your book. They also can best edit and produce your book and market it to the right audience. For scholarly or technical books in the sciences, university presses and commercial scientific publishers often prove most appropriate. Popular books in the sciences often are served well by commercial publishers that include such books among their specialties. Some university presses also excel at publishing science books for general readerships.

Whether the idea for the book is yours or a publisher's, you generally must submit a proposal before receiving approval to prepare and submit the manuscript. Typically, the proposal includes an annotated table of contents, a description of the intended market for the book, a sample chapter, and curriculum vitae or resume. To help decide whether to accept the proposal, the publisher may send it out for peer review. The publisher also will do a financial analysis; if the expected profits do not seem to justify the cost of producing the book, the publisher may decline the project even if it seems otherwise promising. Sometimes, however, another, perhaps more specialized publisher will then accept the project. For example, sometimes a university press but not a commercial publisher agrees to publish a book that is of scientific importance but for which sales are expected to be low.

Book proposals, unlike scientific papers, may be submitted to more than one publisher at once. If, however, a proposal is being submitted simultaneously, the author should inform the publishers. For specialized scientific books, the author typically submits a proposal directly to the publisher. If, however, a book seems likely to sell very well, using an agent can be advisable.

If a proposal is accepted, the publisher is likely to offer the author an advance contract to sign. (Some publishers, however, do not generally offer a contract until the book manuscript is completed and accepted.) An advance contract, which typically runs several pages, usually specifies such items as length, maximum number of figures and tables, deadline, royalties paid to the author, electronic rights, and even film rights (not a likely concern for most book authors

"I'm not sure, but I did notice that there was a letter from a publisher in his box."

(© Vivian S. Hixson, reproduced by permission)

in the sciences). Review the contract carefully. If modifications seem called for, work with the publisher to come to an agreement.

An advance contract is not a guarantee that the book will be published. It does indicate, however, that if you satisfactorily complete the manuscript, publication should proceed. In the sciences, unlike in fiction writing, you generally should have a contract before doing most of the work on a book.

HOW TO PREPARE A BOOK MANUSCRIPT

Joy at signing a book contract can readily become terror as the prospect looms of writing several hundred manuscript pages. Breaking the project into manageable chunks, however, can keep it from becoming overwhelming. While still remembering the scope of the book, focus on one chapter, or part of a chapter, at a time. Soon you might be amazed at how much you have written. Unless chapters build directly on each other, you may be able to write them in whatever order you find easiest. Similarly, a chapter, like a scientific paper, often need not be written from start to finish.

Much as journals have instructions for authors, book publishers have author guidelines. These guidelines, which sometimes can be accessed from publishers' websites, present the publisher's requirements or preferences regarding manuscript format, preparation of tables and figures, and other items, such as obtaining permission to reprint copyrighted materials. The guidelines also may specify the style manual to follow. Before starting to write, look carefully at the guidelines. For convenience, perhaps prepare a sheet listing the main points to remember about the manuscript format, print it on colored paper for easy identification, and place it where you readily can consult it. Following the instructions can save you, and the publisher, effort later.

Immediate demands on your time can easily rob you of opportunity to work on a book. If possible, set aside specific times for writing. For example, include in your regular weekly calendar some blocks of time to work on the book, as if they were appointments. Or have certain times of the year to focus on writing. If opportunity permits, perhaps arrange beforehand to get a sabbatical leave to work on the book, or negotiate to have reduced duties while doing so.

For a busy scientist-author, the writing of a book can extend over months or years, sometimes with interruptions of weeks or more. Therefore, a consistent style and voice can be difficult to maintain. One tactic that can help address this problem: Before resuming your writing, reread, or spend a little time editing, a section you have already drafted. Also, once you have drafted the entire book manuscript and are revising it, look for consistency.

And yes, be prepared to revise the manuscript. In books, as in scientific papers, good writing tends to be much-revised writing. Some book authors do much of the revising as they go, a paragraph or subchapter at a time, and then have little more than final polishing left. Others do a rough draft of the entire manuscript and then go back and refine it. Take whatever approach works for you. But one way or another, do revise.

If the book will include material for which you do not hold copyright—for example, illustrations published elsewhere—you will need permission unless the material is in the public domain. You also may need to pay permission fees. Obtaining the needed permissions is your responsibility, not your publisher's. However, your publisher may be able to provide advice in this regard, and publishers' guidelines for authors often include sample letters for seeking permission. Start the permissions process early; identifying copyright holders, receiving permissions, and (if needed) obtaining images suitable for reproduction sometimes takes many weeks.

Once you submit your book manuscript, the publisher may send it for peer review. Beforehand or simultaneously, consider obtaining peer review of your own. Show the manuscript to people whose views you regard highly, including experts on your subject and individuals representative of the intended readers of your book. Solicit and consider their frank feedback. If appropriate, thank your reviewers in the acknowledgments (with their permission) and give them copies of the book when it appears.

HOW TO PARTICIPATE IN THE PUBLICATION PROCESS

At the publishing company, the proposal for a new book typically goes to an editor in charge of obtaining new manuscripts in your field. This editor, often called an *acquisitions editor,* oversees the review of your proposal, answers questions you may have while preparing the manuscript, and supervises the review of your manuscript. Once your manuscript is accepted, responsibility commonly moves to another editor, sometimes called a *production editor,* who coordinates the editing of the manuscript and other aspects of the conversion of your manuscript into a book.

Open communication with the editors facilitates publication. If, as you prepare the manuscript, you have questions about format, permissions, potential changes in content, or other matters, ask the acquisitions editor. Getting the answer now may save much time later. If you fall behind and might not be able to meet deadlines, inform the acquisitions editor promptly, so that, if necessary, plans can be revised. Similarly, if at times during the editing and production phase you will not be available to review materials or answer questions, inform the production editor so that schedules can be designed or adjusted accordingly.

Book manuscripts in the sciences, like scientific papers, commonly undergo peer review. Your editor may do a preliminary assessment to determine whether the manuscript is ready for peer review or whether revisions are needed first. Once the manuscript is ready for peer review, you may be able to help the editor by suggesting experts in your field to consider including among the reviewers. After peer review is complete, the publisher will decide how to proceed. At a university press, a committee of faculty members is likely to advise the publisher in this regard.

For a book manuscript, as for a scientific paper, any of four decisions may be reached. Commonly, the manuscript will be accepted but some revisions will be required. Occasionally, the manuscript will be accepted without revisions. Sometimes, if the manuscript needs major revision, the author will be asked to revise it and submit it for reevaluation. And sometimes, if a manuscript has fallen far short of its seeming potential, it will not be accepted.

In the likely instance that some revisions are required, the editor will indicate how to proceed. Commonly, you will receive peer reviewers' suggestions. You also should receive guidance from the editor—for example, regarding which suggestions are important to follow and which are optional, or what to do about contradictory advice from different reviewers. The editor will also determine with you a timetable for completing the revisions.

Once your manuscript is successfully revised, the book will enter production. In this phase, a copy editor will edit the manuscript, a designer will design the book, and ultimately, the book will go to the printer.

Your manuscript probably will go to a freelance copy editor who specializes in editing book manuscripts in your field. Because this copy editor knows your field and the conventions in it, he or she can edit your manuscript more appropriately than a general copy editor could. Your communication with the copy editor is likely to be through the production editor coordinating publication of your book. You will receive the edited manuscript for review, as well as any *queries* (questions) the copy editor might have, for example about points that seem inconsistent or otherwise in need of clarification. Check the edited manuscript in the time allotted; if inaccuracies or other problems have been introduced, correct them. Answer any queries so that necessary changes can be made.

In addition to receiving the edited manuscript to check, you will later receive page proofs—that is, copies of the draft pages of the book. Review the page proofs promptly but thoroughly. Make sure that nothing has been omitted, that all corrections of the edited manuscript were entered accurately, and that all photographs and other illustrations are included and correctly oriented. Limit your changes, however, to those that are necessary. Now is not the time for rewriting.

For many books in the sciences, a good index is crucial. Once the page proofs are ready and thus one can see what information will appear on what page, an index can be prepared. Some authors prepare the indexes for their books themselves. Others, however, use professional indexers. Indexing is a highly skilled craft, and often a professional indexer can prepare a more useful index than the author could. A professional indexer also is likely to prepare the index more efficiently. If your book will be professionally indexed, your publisher should be able to identify and hire a suitably qualified indexer. In some cases, the contract for your book may indicate that the publisher will pay for indexing. If you are to pay, the publisher may deduct the sum from your book royalties rather than ask you to pay directly. In any case, the money is likely to be well spent.

HOW TO HELP MARKET YOUR BOOK

If you have chosen well, your publisher will have experience and expertise marketing books to the audiences for yours. To do its best job, though, the publisher needs information from you. Thus, you are likely to receive an *author questionnaire*. The questionnaire may, for example, ask you to identify scientific organizations that have members interested in your topic, conferences at which your book might appropriately be sold, journals for which your book is suitable for review, and people well suited to provide endorsements. The questionnaire also is likely to request information about you, as well as other information that can aid in promoting your book. Take the time to complete the questionnaire thoroughly. The information can help the marketing department ensure that the appropriate audience knows of your book and thus that your book receives the sales it deserves.

Increasingly, authors are expected to take active roles in marketing, especially if their books are for general audiences. "Today, I do not offer a contract or invest in a project if the author isn't willing to promote his or her book," states an acquisition editor at a university press. "In today's world of Facebook, YouTube, Twitter, blogs, and so on, we ask authors to actively seek venues in which to speak, lecture, present—anything to get the book into the right hands. Markets are increasingly specialized and targeted, and a reader is more likely to purchase a book on astronomy at, say, a star party where the author is a featured guest than by walking into a Barnes and Noble and reaching for that book among the other 150,000 or so titles available each year."

You may also be asked to participate in the marketing of a book in other ways. For example, you may be interviewed for radio or television or for a podcast or webinar. Book signings may be arranged. Arrangements may be made for excerpts of the book to appear in magazines. Be open to such possibilities, and suggest any that occur to you. If you have questions, consult the marketing department.

For scholarly or technical books, marketing remains more restrained. Although pushing one's book in inappropriate venues, such as scientific presentations, can be counterproductive, do mention your book when suitable occasions arise. For example, if a posting in an email discussion list requests information that your book happens to contain, mention your book. Likewise, consider mentioning your book in science blogs and on professionally oriented social networking sites. Doing so can at least prompt prospective users to seek the book in the library. And given the ideals of science, the success of a book should be measured not only in sales but also in service to those who can benefit.

CHAPTER 26

How to Write for the Public

Regard readers not as being ignorant but, more likely, innocent of your topic and its jargon. Write for them, not at them.

—Alton Blakeslee

WHY WRITE FOR GENERAL READERSHIPS?

Preparing papers and proposals for peers to read can entail plenty of writing. Why might you write for nonscientists also?

Sometimes your academic program or job includes doing some writing for lay readerships. For example, requirements for a graduate degree can include writing a nontechnical summary of your thesis. Your funding agency may require public outreach. Or, if you teach introductory courses in your discipline, you may prepare teaching materials that are essentially for the public.

At our own initiative as well, some of us write for the public. Some of us enjoy doing such writing and appreciate the chance to reach audiences broader than those in our own fields. Other motivations can include giving members of the public useful information on technical topics, helping to attract people to scientific careers, and helping to engender public support for science. Some of us also welcome the bit of extra income that popular writing can bring.

FINDING PUBLICATION VENUES

If you wish to write for the public, how might you find a home for your work? Good places to start can be publications, both online and in print, that you like to read. Do not limit yourself to those devoted solely to science. Other publications, including magazines focusing on specific interests or geared to specific population groups, often contain articles on science-related topics. If you have not published articles for the public before, suitable starting points can include local, regional, or specialized publications, including those at your own institution. Another good starting point can be a blog that you establish or an existing blog for which you arrange to write guest posts. Then, once you have proven your ability to write for the public, publications of greater scope are more likely to welcome your requests to write.

If a venue seems suitable, try to determine whether it accepts freelance work. One way is to see who writes for it. If all the authors are staff members listed in its masthead, a magazine is unlikely to accept your work. But if, for instance, some articles have blurbs saying that they are by scientists, the venue might be appropriate for you.

Many popular publications that accept freelance work have writer's guidelines, which are analogous to journals' instructions to authors. Look for these guidelines, which appear on the publications' websites or can be obtained from their editorial offices. Items often addressed include subject areas in which articles are wanted (and not wanted), standard article lengths, requested writing style, rates of payment, and postal or electronic addresses to which article proposals should be submitted.

Typically, magazines want prospective authors to submit article proposals, known as *query letters,* rather than submitting completed articles at the outset. Doing so is more efficient for the author, who can thus avoid wasting time writing articles that the magazine would not want. It also is more efficient for the editor: By reading a query letter, the editor can quickly evaluate the story idea and the writer's skill. And if the query is accepted, the editor can work with the writer from the outset to suit the story to the magazine's needs.

A query letter generally should be limited to one page (or the equivalent amount of text in an email message). Begin by describing the article you propose. Among questions you might address are the following: What is the main topic of the article, and what major subtopics do you plan to address? Why is the topic likely to interest readers? What information sources do you expect to use? How might the article be organized? What types of photographs or other graphics might be appropriate? Near the end of the letter, include a paragraph summarizing your qualifications to write the article. If you have not written for the magazine before, provide examples, if available, of articles you have written for the public. Further information on writing query letters, and more generally on writing for magazines, can be found in books such as *You Can Write for Magazines* (Daugherty 1999) and *The Complete Guide to Article Writing* (Saleh 2013) and in magazines such as *Writer's Digest.*

Before writing for a magazine, website, or other venue, analyze writing that it has published or posted, so that yours can fit in. Notice, for example, how long the paragraphs tend to be, how formal or informal the wording is, whether headings divide the articles into sections, and whether articles tend to include bulleted lists. In writing for a popular venue, as in writing a scientific paper, suiting the writing to the site will increase likelihood of publication.

ENGAGING THE AUDIENCE

Readers of journals where your papers appear are likely to be interested already in your topic. Or at least they are deeply interested in science. Thus, beyond perhaps noting the importance of the topic, you generally need to do little to attract readers.

When writing for the public, however, you typically must do more to engage the audience. One key to engaging the audience is analyzing the audience. The public is not uniform. Rather, readers of different publications have different interests. Ditto for users of different websites and audiences of different broadcast programs. Consider what the audience members are likely to care about, and relate what you say to those interests.

Regardless of other interests, most people care about people. Thus, use human interest to help engage the audience. For example, tell about the people who did the research. If there are technology users or patients, tell about them as well. When appropriate, also include almost-human interest, for much of the public likes animals.

Include quotes from the people in your piece. Doing so contributes to human interest and can keep attention through varied voices and lively wording. To obtain quotes, of course, you generally must do interviews even if you are well versed on the topic about which you are writing.

People generally like stories, which often combine human interest and suspense. So consider including some narrative. For example, show how a line of research developed—and do not omit the difficulties encountered. Or include some anecdotes illustrating your points.

Especially with regard to technologies, costs may interest and be important to the public; consider including economic context. Likewise, if relevant to your subject, provide social and ethical context.

Science is full of wonder as well. Use it to help engage the audience. Draw on the audience's curiosity. Too much gee-whiz can cheapen science, but a little can enliven a piece.

In a popular article, unlike in a scientific paper, you may be able to engage in wordplay and other humor. If, for example, puns are your passion, now may be your chance. Be sure, however, that any humor would be understandable to the audience; avoid scientific in-jokes.

Think visually as well as verbally. Editors of popular pieces for print, the web, and television generally want to use photos or other graphics. Even radio stories benefit from description of visual aspects. If a piece is to include visuals, the editor can tell you whether to provide them yourself or merely provide ideas.

To maintain interest, pace the article carefully. Think of a popular article as a chocolate chip cookie. Just as each bite of the cookie should contain at least one chocolate chip, each few paragraphs of the article should contain something tasty—for example, a good quote, a lively anecdote, or a deft analogy. Keep your readers wanting one more bite.

CONVEYING CONTENT CLEARLY

Much of what you do to engage the audience also can aid in conveying content clearly. For example, gearing your piece to the audience, using lucid analogies, and providing visuals can serve both roles. So can supporting what you say with examples.

Members of the public probably will not know technical terms in your field. Where feasible, avoid such jargon. If technical terms are important to the story you are telling, or if readers should learn them for future use, remember to define them. One way to avoid intimidating readers is to state an item in familiar words before providing the technical term (example: "bone-forming cells called osteoblasts"). Remember also to define abbreviations. "PCR" may be everyday language for you but meaningless to your readers.

Structure what you say to promote clarity. For instance, provide overviews before details. Explicitly state the relationships between concepts. Repeat important points.

Include numbers; members of the public often expect and enjoy them. However, present them in easily understood ways. If the audience is unfamiliar with metric units, use English units. And relate sizes to familiar ones ("about the size of . . ."). Do not overwhelm readers with many numbers clustered together. Separate pieces of "hard stuff" with softer material, such as anecdotes and examples.

Sometimes readers have misconceptions about scientific items. To counter misconceptions without seeming condescending, consider taking the following approach (Rowan 1990): First, state the commonly held view and note its seeming plausibility. Then show the inadequacy of that view. Finally, present the scientifically supported view and explain its greater adequacy.

Of course, follow the principles of readable writing presented elsewhere in this book. For example, use concise, straightforward language when possible. Structure sentences simply. Avoid lengthy paragraphs.

Finally, consider checking with readers. Show a draft to nonscientist friends or neighbors or family members. See what they find interesting. See what they find clear or unclear. Then consider revising your piece accordingly before submitting it.

EMULATING THE BEST

Further guidance in writing for the public about science appears in a variety of books and articles (for example, Blakeslee 1994; Blum, Knudson, and Henig 2006; Gastel 1983, 2005; Hancock 2003; Stocking et al. 2011; Writers of Sci-Lance 2013).

In addition, good popular science writing, like good writing for scientific audiences, benefits from following good examples. Where can you find such examples? Major newspapers and magazines contain much good science writing. So do the bestseller lists. Fine pieces of popular science communication in various media have won AAAS Kavli Science Journalism Awards, National Association of Science Writers Science in Society Journalism Awards, and Pulitzer Prizes; the websites for such awards list recipients and, in some cases, include links to the pieces. Other sources of excellent examples include the annual anthology titled *The Best American Science and Nature Writing*. Consume good works of popular science communication. Whether or not you explicitly analyze them, you are likely to assimilate much about writing skillfully for the public.

PART VI

Conference Communications

CHAPTER 27

How to Present a Paper Orally

Talk low, talk slow, and don't say too much.

—John Wayne

HOW TO GET TO PRESENT A PAPER

The first step in presenting a paper is to obtain a chance to do so. Sometimes, you might receive an unsolicited invitation. For major conferences, however, you normally must take the initiative by submitting an abstract of the paper that you hope to present.

Those organizing the conference typically provide abstract submission forms; these usually can be accessed and submitted via the web. The submitted abstracts undergo peer review, and the submitters whose abstracts seem to describe the strongest research are asked to give oral presentations. For some conferences, those whose abstracts represent good work of lower priority are asked to give poster presentations. For other conferences, separate application processes exist for oral presentations and for posters.

Those who decide whether you should present a paper are likely to have only your abstract on which to base their decision. Therefore, prepare the abstract carefully, following all instructions. Word the abstract concisely, so it can be highly informative although it must be brief. (The word limit sometimes is higher than that for abstracts accompanying published papers, but be sure to stay within it.) If figures or tables are allowed, follow all instructions, and do not exceed the number permitted. Organize the abstract well—typically in the same sequence as a scientific paper. Also write clearly and readably, as those reviewing the abstracts probably will be busy scientists with many abstracts to review and little patience with those that are unclear on first reading. Of course, be sure to submit the abstract by the deadline. Present your research well in your abstract, and you may soon be presenting a paper.

For many conferences, the peer reviewers might not be the only ones seeing your abstract. Often, presentation abstracts are printed in the conference program, posted on the conference website, or both. Those reading them can include conference registrants trying to decide which sessions to attend, fellow scientists unable to attend the conference but interested in the content, and science reporters trying to determine which sessions to cover. All the more reason to provide an informative and readable abstract.

A WORD OF CAUTION

If you receive an unsolicited invitation to speak at a conference that you have not have heard of, check into the matter carefully rather than automatically accepting. In recent years, what are known as *predatory conferences* have arisen. These are not valid scientific conferences but rather scams to take people's money. The organizers invite prospective attendees, obtain their advance registration fees, and then either hold a conference with little scientific substance or hold no conference at all.

One clue that a conference might be predatory is an invitation to speak at a conference that is outside your field. Other possible clues include an invitation that emphasizes the beautiful location rather than the conference content, lists fees that are much higher than usual, or has many grammatical errors and misspellings. If you are early in your career, perhaps consult a mentor or senior colleague to help determine whether a conference is valid. Also, online searching can help identify conference organizers that credible sources, such academic librarians, have deemed questionable.

ORGANIZATION OF THE PAPER

The best way to organize a paper for oral presentation generally is to proceed in the same logical pathway that one usually does in writing a paper, starting with "What was the problem?" and ending with "What is the solution?" However, it is important to remember that oral presentation of a paper does *not* constitute publication, and therefore different rules apply. The greatest distinction is that the published paper must contain the full experimental protocol, so that the experiments can be repeated. The oral presentation, however, need not and should not contain all of the experimental detail, unless by chance you have been called on to administer a soporific at a meeting of insomniacs. Extensive citation of the literature is also undesirable in an oral presentation.

PRESENTATION OF THE PAPER

Most oral presentations are short (with a limit of 10 minutes at many meetings). Thus, even the theoretical content must be trimmed down relative to that of a written paper. No matter how well organized, too many ideas presented too quickly will be confusing. You should stick to your most important point or result and stress that. There will not be time for you to present all your other neat ideas.

There are, of course, other and longer types of oral presentations. A typical time allotted for symposium presentations is 20 minutes. A few are longer. A seminar is normally 1 hour. Obviously, you can present more material if you have more time. Even so, you should go slowly, carefully presenting a few main points or themes. If you proceed too fast, especially at the beginning, your audience will lose the thread; the daydreams will begin, and your message will be lost.

Time limits for conference presentations tend to be strictly enforced. Therefore carefully plan your presentation to fit the allotted time—lest you be whisked from the podium before you can report your major result. If possible, make your presentation a bit short (say, 9 or 9.5 minutes if 10 minutes are allotted), to accommodate unexpected slowdowns. Rehearse your presentation beforehand, both to make sure it is the right length and to help ensure smooth delivery. During your presentation, stay aware of the time. Perhaps indicate in your notes what point in the presentation you should have reached by what time, so that if necessary you can adjust your pace.

A few more pointers on delivery: Speak very clearly, and avoid speaking quickly, especially if the language in which you are presenting is not the native language of all the audience members. Remember to look at the audience. Show interest in your subject. Avoid habits that might be distracting—such as jangling the change in your pocket or repeatedly saying "um" or "you know" or the equivalent from your native language. To polish your delivery, consider videoing rehearsals of one or more of your presentations.

Does stage fright plague you? Consider the following suggestions: Prepare well so you can feel confident, but do not prepare so much that you feel obsessed. To dissipate nervous energy, take a walk or take advantage of the exercise facilities in the conference hotel. Beware of too much caffeine, food, or water. Hide physical signs of anxiety; for example, if your hands tremble under stress, do not hold a laser pointer. Realize that a presentation need not be flawless to be excellent. Perhaps most important, realize that the audience members are there not because they wish to judge your speaking style but because they are interested in your research.

(www.phdcomics.com)

SLIDES

At small, informal scientific meetings, various types of visual aids—including flip charts, whiteboards, and blackboards—may be used. At most scientific conferences, however, PowerPoint presentations are the norm. Every scientist *should* know how to prepare effective slides and use them well, yet attendance at almost any meeting quickly indicates that many do not.

Here are a few important considerations. First, slides should be designed specifically for use with oral presentations, with large enough lettering to be seen from the back of the room. In general, use lettering that is at least 28 points in size. Choose a sans serif typeface, such as Arial or Calibri. Slides prepared from graphs that were drawn for journal publication are seldom effective and often are not even legible. Slides prepared from a printed journal or book are almost never effective.

Slides should be uncrowded. Each slide should illustrate a particular point or perhaps summarize a few. To permit rapid reading, use bullet points, not paragraphs. For text slides, try not to exceed about seven lines of about seven words each—or, stated another way, about 50 words in total. It has been said that if a slide cannot be understood in 4 seconds, it is a bad slide.

Beware of showing too many slides. A moderate number of well-chosen slides will enhance your presentation; too many will be distracting. One general guideline is not to exceed an average of about one slide per minute. If you show a slide of an illustration or table, indicate its main message. As one long-suffering audience member said, "Don't just point at it."

Speaking of illustrations and tables: If there are findings that you can present in either a graph or a table, use a graph in an oral presentation. Doing so will help the audience grasp the point more quickly. And speaking of pointing:

If you use a laser pointer, take care with it. In your enthusiasm or distraction, do not wildly gesture at the slide—or the audience—with a lighted pointer. Rather, turn on the laser pointer only when you want to call attention to a specific item on a slide. Direct the laser pointer specifically at the item. And if, for example, you are showing a pathway, trace it with the pointer. If you shake during presentations, hold the laser pointer in one hand, and use the other hand to steady that hand.

If the conference has a speaker ready room (a room in which speakers can test their audiovisuals), check that your slides are functioning properly. Also, if possible, get to the hall before the audience does. Make sure the projector is working, ascertain that your slides will indeed project, and check the lights. If you will use a microphone, ensure it is functioning.

Normally, each slide should make one simple, easily understood visual statement. The slide should supplement what you are saying when the slide is on the screen; it should *not* simply repeat what you are saying. Except when doing so could help overcome a language barrier, do not read the slide text to the audience.

A nice touch, and a tradition in some research areas, is to include a closing slide acknowledging collaborators and perhaps showing a photo of the research group. If the research being reported was a team effort, consider including such a slide if appropriate in your field.

Slides that are thoughtfully designed, well prepared, and skillfully used can greatly enhance the value of a scientific presentation. Poor slides would have ruined Cicero.

THE AUDIENCE

The presentation of a paper at a scientific meeting is a two-way process. Because the material being communicated at a scientific conference is likely to be the newest available information in that field, both the speakers and the audience should accept certain obligations. As indicated earlier in this chapter, speakers should present their material clearly and effectively so that the audience can understand and learn from the information being communicated.

Almost certainly, the audience for an oral presentation will be more diverse than the readership of a scientific paper. Therefore, the oral presentation should be pitched at a more general level than would be a written paper. Avoid technical detail. Define terms. Beware of using acronyms the audience does not already know. Explain difficult concepts. Repeat important points.

Rehearsing a paper before the members (even just a few members) of one's own department or group can make the difference between success and disaster.

For communication to be effective, the audience also has various responsibilities. These start with simple courtesy. The audience should be quiet and attentive, no matter how compellingly a mobile device may beckon. Speakers respond well to an interested, attentive audience, whereas the communication process can be virtually destroyed when the audience is noisy, distracted, or, worse, asleep.

The best part of an oral presentation is often the question-and-answer period. During this time, members of the audience have the option, if not the obligation, of raising questions not covered by the speaker, and of briefly presenting ideas or data that confirm or contrast with those presented by the speaker. Such questions and comments should be stated courteously and professionally. This is not the time (although we have all seen it) for some windbag to vent spleen or to describe his or her own erudition in infinite detail. It is all right to disagree, but do not be disagreeable. In short, the speaker has an obligation to be considerate to the audience, and the audience has an obligation to be considerate to the speaker.

A FEW ANSWERS ON QUESTIONS

What should you do if an audience member is indeed abrasive? If someone asks an irrelevant question? If a question is relevant but you lack the answer?

If someone is rude, stay calm and courteous. Thank him or her for the question or comment, and if you have a substantive reply, provide it. If the person keeps pursuing the point, offer to talk after the session.

If a question is irrelevant, take a cue from politicians and try to deflect the discussion to something related that you wish to address—perhaps a point you had hoped to include in your presentation but lacked time for. ("That's an interesting question, but a more immediate concern to us was. . . .") Alternatively, offer to talk later.

If you lack the answer to a question, do not panic—and definitely do not bluff. Admit that you do not know. If you can provide the answer later, offer to do so; if you know how to find the answer, say how. To help prepare for questions that might arise, have colleagues quiz you after you rehearse.

Especially if you have not yet submitted for publication the work you are presenting, consider making note of the questions and comments (or having a colleague do so). Audience members can function as some of your earliest peer reviewers. Keeping their questions in mind when you write may strengthen your paper and hasten its acceptance.

CHAPTER 28

How to Prepare a Poster

It takes intelligence, even brilliance, to condense and focus information into a clear, simple presentation that will be read and remembered. Ignorance and arrogance are shown in a crowded, complicated, hard-to-read poster.

—Mary Helen Briscoe

POPULARITY OF POSTERS

In recent decades, posters presenting research have become ever more common at national and international meetings. Sessions featuring such posters originated—apparently in the late 1960s through mid-1970s (Waquet 2008)—as follows: As attendance at meetings increased, and as pressure mounted on program committees to schedule more and more papers for oral presentation, something had to change. The large annual meetings, such as those of the Federation of American Societies for Experimental Biology, reached the point where the number of available meeting rooms no longer sufficed. And, even when sufficient numbers of rooms were available, the resulting large numbers of concurrent sessions made it difficult or impossible for attending scientists to keep up with the work being presented.

At first, program committees simply rejected whatever number of abstracts was deemed to be beyond the capabilities of meeting room space. Then, as poster sessions were developed, program committees were able to take the sting out of rejection by advising the "rejectees" that they could consider presenting their work as posters. In the early days, the posters were relegated to the hallways of the meeting hotels or conference centers; nevertheless, many authors, especially graduate students trying to present their first paper, were happy to have their work accepted for a poster session rather than being knocked off the program entirely. Also, the younger generation of scientists had come of age during the era of science fairs, and they liked posters.

Nowadays poster sessions have become an accepted and meaningful part of many meetings. Large societies set aside substantial space for the poster presentations; at some meetings, thousands of posters are presented. Even small societies often encourage poster presentations, because some types of material may be presented more effectively in poster graphics and the accompanying discussion than in the confines of the traditional 10-minute oral presentation.

Meanwhile, posters and poster sessions continue to evolve. Recent developments include *electronic posters,* also known as *e-posters* or *digital posters.* More and more conferences feature e-poster sessions, for which posters are provided digitally and displayed electronically. Some such sessions are limited to e-posters that have only static images and so are largely digital equivalents of conventional posters. Others display e-posters that incorporate dynamic elements such as videos and animations. Also, some conferences include related sessions such as sets of 3-minute spoken "flash poster presentations" intended to interest attendees in visiting the speakers' posters.

As poster sessions have become larger and more complex, the rules governing the preparation of posters have become much stricter. When many posters must be fitted into a given space, obviously the requirements have to be carefully stated.

Before starting to prepare a poster, be sure to know the requirements specified by the meeting organizers. You of course must know the height and width of space available. The minimum sizes of type may be specified, as may other aspects, such as requirements for e-posters. As well as being given to the presenters, this information is likely to be available on the conference website.

ORGANIZATION

The organization of a poster normally should follow the IMRAD format (introduction, methods, results, and discussion), although graphic considerations and the need for simplicity should be kept in mind. There is very little *text* in a well-designed poster, most of the space being used for illustrations. In general, a poster should contain no more than 500 to 1,000 words (approximately the number of words in two to four double-spaced pages of a manuscript or in two to four typical article abstracts). If a poster is in landscape format, with the width exceeding the height, placing the content in three to five vertical columns generally works well. For posters in portrait format, two or three columns is usually the best choice. Unless the conference organizers require an abstract on your poster, do not include one; the poster as a whole is not much more extensive than an abstract, so an abstract tends to be redundant and waste valuable space. Where feasible, use bulleted or numbered lists rather than paragraphs. If paragraphs are used, keep them short, for readability.

The introduction should present the problem succinctly; the poster will fail unless it has a clear statement of purpose at the beginning. The methods section will be very brief; sometimes a sentence or two will suffice to describe the type of approach used. The results section, which is often the shortest part of a written paper, is usually the major part of a well-designed poster. Most of the available space should be used to illustrate results. The discussion should be brief. Some of the best posters do not even use the heading "Discussion"; instead, the heading "Conclusions" appears over the far-right panel, the individual conclusions perhaps being in the form of numbered or bulleted short sentences. Literature citations should be kept to a minimum.

PREPARING THE POSTER

Preparing a poster often begins with preparing an abstract for the selection committee. Like that for an oral presentation, this abstract should be carefully written. It should conform to all stated requirements, and it should be readably worded, for easy peer review. Before writing the abstract, think ahead to what the poster would look like. Choose as your topic a part of your research that is focused enough to present effectively as a poster (Mitrany 2005) rather than trying to cover so much that a bafflingly cluttered poster would result.

You should number your poster to agree with the program of the meeting. The title should be short and (if feasible) attention-grabbing; if it is too long, it might not fit. The title should be readable out to a distance of 10 feet (about 3 m). The typeface should be bold and dark, and the type should be at least about 1 inch (about 25 mm) high—in other words, at least about 72 points. Unless the conference organizers require titles to be in capital letters, use mainly lowercase letters; in addition to taking up less space, they make the title easier to read as lowercase letters vary more in shape than capital letters do. (Compare "PRESENTING POSTERS" and "Presenting Posters.") The names of the authors should be somewhat smaller. The text type should be large enough to be readily readable (normally at least 18 points). Large blocks of type should be avoided; where feasible, use bulleted or numbered lists.

A poster should be self-explanatory, allowing different viewers to proceed at their own pace. If the author must spend most of his or her time merely explaining the poster rather than responding to scientific questions, the poster is largely a failure.

Having lots of white space throughout the poster is important. Distracting clutter will drive people off. Try to make it *very* clear what is meant to be looked at first, second, and so forth (although many people will still read the poster backward). Visual impact is particularly critical in a poster session. If you lack graphic talent, consider getting the help of a graphic artist, for example from the media resources department at your institution.

A poster should contain *highlights* so that passersby can easily discern whether the poster is something of interest to them. If they are interested, there will be plenty of time to ask questions about the details. Also, consider preparing handouts containing more detailed information; they will be appreciated by colleagues with similar specialties.

A poster may actually be better than an oral presentation for showing the results of a complex experiment. In a poster, you can organize the highlights of several threads well enough to give informed viewers the chance to recognize what is going on and then get the details if they so desire. The oral presentation, as stated in the preceding chapter, is better for getting across a single result or point.

The really nice thing about posters is the variety of illustrations that can be used. There is no barrier (as there often is in journal publication) to the use of color. All kinds of photographs, graphs, drawings, paintings, radiographs, maps, and even cartoons can be presented. Try to use images that both attract and inform. Make the images large enough to see easily, and keep them simple enough to understand quickly.

Once the poster is drafted, check it carefully. Be sure, for example, that all illustrations are clearly labeled and that the poster includes your contact information. Proofread the poster, and have others do so—lest you discover too late that your coinvestigator's name was misspelled. If you are traveling by airplane to the conference, carry your poster with you. Do not check it in your luggage—which might be delayed until after the poster session if, as happened to a colleague of ours, you are flying to San Jose but your luggage gets routed to San Juan. Regardless of whether your poster is conventional or electronic, have a backup copy is case the original is lost, destroyed, or damaged. For example, carry a copy on a USB drive, email a copy to yourself, save a copy in the cloud—or do more than one of these.

There are many excellent posters. Some scientists do indeed have considerable creative ability. It is obvious that these people are proud of the science they are doing and that they are pleased to put it all into a pretty picture.

There are also many terrible posters. A few are simply badly designed. The great majority of *bad* posters are bad because the author is trying to present too much. Huge blocks of typed material, especially if the type is small, will not be read. Crowds will gather around the simple, well-illustrated posters; the cluttered, wordy posters will be ignored.

PRESENTING THE POSTER

A poster presentation is, as its name says, both poster and presentation. Typically, for some of the time the poster is on display, one or more of the authors accompany and discuss it. Thus, preparing a well-designed poster constitutes only part of a successful poster presentation.

Leave your shyness behind when you accompany a poster. Now is not the time to hide behind the poster or stare at your shoes. Think ahead about questions you might be asked, and verbally and otherwise show a readiness to answer questions. If occasion arises, ask questions as well. Take advantage of the chance for feedback. Also take advantage of the chance to network. Those talking with you might well include potential collaborators or employers.

What should you wear when presenting a poster? At some conferences, poster presenters typically wear suits. At others, they usually dress more casually. If in doubt, ask a mentor or colleague who knows the norms. One lighthearted report of a very small study (Keegan and Bannister 2003) suggests that wearing colors that coordinate with those of a poster might increase the number of visitors to the poster. A photo of a presenter wearing clothes color-coordinated with his poster appears on the web page "Designing Conference Posters" (colinpurrington.com/tips/poster-design), which presents extensive advice on preparing and presenting posters; scroll down patiently to find this photo, for this web page is extensive.

As noted, consider having handouts available that present your work in more detail; remember to include your contact information. Also consider having printouts of your poster and copies of papers describing related research you have done. If you run out of handout materials or wish to share materials that you did not bring, obtain email addresses and send the materials as attachments. Perhaps have business cards available too. And if, for example, you are seeking a postdoctoral fellowship or a job, perhaps have copies of your curriculum vitae or résumé on hand.

In short, take advantage of the interactive opportunities of the poster session. As your professional community comes to you, present your work and yourself at your best.

CHAPTER 29

How to Write a Conference Report

Conference: a gathering of important people who singly can do nothing, but together decide that nothing can be done.

—Fred Allen

DEFINITION

A conference report can be one of many kinds. However, let us make a few assumptions and, from these, try to devise a picture of what a more-or-less typical conference report should look like.

It all starts, of course, when you are invited to participate in a conference (congress, symposium, workshop, panel discussion, seminar, colloquium), the proceedings of which will be published. At that early time, you should stop to ask yourself, and the conference convener or editor, exactly what is involved with the publication.

The biggest question, yet one that is often left cloudy, is whether the proceedings volume will be defined as primary. If you or other participants present previously unpublished data, the question arises (or at least it should) as to whether data published in the proceedings have been validly published, thus precluding later republication in a primary journal.

The clear trend, it seems, is to define conference reports as not validly published primary data. This is seemingly in recognition of three important considerations: (1) Most conference proceedings are one-shot, ephemeral publications, not purchased widely by science libraries around the world; thus, because of limited circulation and availability, they fail one of the fundamental tests of valid publication. (2) Most conference reports either are essentially review papers, which do not qualify as primary publication, or are preliminary reports presenting data and concepts that may still be tentative or inconclusive and that the scientist would not yet dare to contribute to a primary publication. (3) Conference reports are normally not subjected to peer review or to more than minimal editing; therefore, because of the lack of any real quality control, many reputable publishers now define proceedings volumes as nonprimary. (There are of course exceptions. Some conference proceedings are rigorously edited, and their prestige is the equal of primary journals. Indeed, some conference proceedings appear as issues of journals.)

This is important to you because you can determine whether your data will be buried in an obscure proceedings volume. It also answers in large measure how you should write the report. If the proceedings volume is adjudged to be primary, you should (and the editor will no doubt so indicate) prepare your manuscript in journal style. You should give full experimental detail, and you should present both your data and your discussion of the data as circumspectly as you would in a prestigious journal.

If, on the other hand, you are contributing to a proceedings volume that is not a primary publication, your style of writing may be (and should be) quite different. The fundamental requirement of reproducibility, inherent in a primary publication, may now be ignored. You need not, and probably should not, have a materials and methods section. Certainly, you need not provide the intricate detail that might be required for a peer to reproduce the experiments.

Nor is it necessary to provide the usual literature review. Your later journal article will carefully fit your results into the preexisting fabric of science; your conference report should be designed to give the news and the speculation for today's audience. Only the primary journal need serve as the official repository.

FORMAT

If your conference report is not a primary scientific paper, just how should it differ from the usual scientific paper?

A conference report is often limited to one or two printed pages, or 1,000 to 2,000 words. Commonly, authors are provided with a simple formula, such as "up to five manuscript pages, double-spaced, and not more than three illustrations (any combination of tables, graphs, or photographs)."

Today, conference reports often appear in electronic formats, either instead of or in addition to print. However, the principles remain the same.

PRESENTING THE NEW IDEAS

As stated, the conference report can be relatively short because most of the experimental detail and much of the literature review can be eliminated. In addition, the results can usually be presented in brief form. Because the full results will be presumably published later in a primary journal, only the highlights need be presented in the conference report.

On the other hand, the conference report might give greater space to *speculation*. Editors of primary journals can get quite nervous about discussion of theories and possibilities that are not thoroughly buttressed by the data. The conference report, however, should serve the purpose of the true preliminary report; it should present and encourage speculation, alternative theories, and suggestions for future research.

Conferences themselves can be exciting precisely because they do serve as the forum for presentation of the very newest ideas. If the ideas are truly new, they are not yet fully tested. They may not hold water. Therefore, the typical scientific conference should be designed as a sounding board, and the published proceedings should reflect that ambience. The strict controls of stern editors and peer review are fine for the primary journal but are out of place for the conference literature.

Because conference reports may interest readers largely because of the newness of the ideas, submit your report promptly. Sometimes, the reports are due before the conference. Other times, they are due shortly afterward, allowing you to add ideas that emerged at the conference. In either case, submit your report by the deadline, so as not to delay publication or posting. If your paper is due shortly after the conference, a good approach can be to draft it before the conference and start revising it during the conference, while discussion of your presentation still is fresh in your mind.

The typical conference report, therefore, need not follow the usual introduction, materials and methods, results, discussion progression that is standard for the primary research paper. Instead, an abbreviated approach may be used. The problem is stated; the methodology used is stated (but not described in detail); and the results are presented briefly, with one, two, or three tables or figures. Then, the meaning of the results is speculated about, often at considerable length. There is likely to be description of related or planned experiments in the author's own laboratory or in the laboratories of colleagues who are currently working on related problems.

EDITING AND PUBLISHING

Finally, it is only necessary to remind you that the editor of the proceedings, usually the convener of the conference, is the sole arbiter of questions relating to manuscript preparation. If the editor has distributed instructions to authors, you should follow them (assuming that you want to be invited to other conferences). You might not have to worry about rejection, since conference reports are seldom rejected; however, if you have agreed to participate in a conference, you should follow whatever rules are established. If all contributors follow the rules, whatever they are, the resultant volume is likely to exhibit reasonable internal consistency and be a credit to all concerned.

PART VII

Scientific Style

CHAPTER 30

Use and Misuse of English

Long words name little things. All big things have little names, such as life and death, peace and war, or dawn, day, night, love, home. Learn to use little words in a big way—It is hard to do. But they say what you mean. When you don't know what you mean, use big words: They often fool little people.

—SSC BOOKNEWS, July 1981

KEEP IT SIMPLE

Earlier chapters of this book outlined the various components that could and perhaps should go into a scientific paper. Perhaps, with this outline, the paper won't quite write itself. But if this outline, this table of organization, is followed, the writing might be much easier than otherwise.

Of course, you still must use the English language if you want your work to have greatest visibility. For some, this may be difficult. If your native language is not English, you may face particular challenges in English-language writing; some suggestions for overcoming those challenges appear in Chapter 34. If your native language is English, you still may have a problem because the native language of many of your readers is not English.

Learn to appreciate, as most manuscript editors have learned to appreciate, the sheer beauty of the simple declarative sentence (subject, then verb, then object). You will thereby avoid most serious grammatical problems and make it easier for people whose native language is not English. You also will make it easier for readers who are busy—as almost all readers of scientific papers are.

DANGLING MODIFIERS

It is not always easy to recognize a dangling participle or related error, but you can avoid many problems by giving proper attention to syntax. The word "syntax" refers to that part of grammar dealing with the way in which words are put together to form phrases, clauses, and sentences.

That is not to say that a well-dangled participle or other misplaced modifier isn't a joy to behold, after you have developed a taste for such things. Those of you who use chromatographic procedures may be interested in a new technique reported in a manuscript submitted to the *Journal of Bacteriology*: "By filtering through Whatman no. 1 filter paper, Smith separated the components."

Of course, such charming grammatical errors are not limited to science. A mystery novel, *Death Has Deep Roots* by Michael Gilbert, contains a particularly sexy misplaced modifier: "He placed at Nap's disposal the marriage bed of his eldest daughter, a knobbed engine of brass and iron."

A Hampshire, England, fire department received a government memorandum seeking statistical information. One of the questions was, "How many people do you employ, broken down by sex?" The fire chief took that question right in stride, answering "None. Our problem here is booze."

If you are interested in harness racing, you might know that the 1970 Hambletonian was won by a horse named Timothy T. According to the *Washington Post* account of the story, Timothy T. evidently has an interesting background: "Timothy T.—sired by Ayres, the 1964 Hambletonian winner with John Simpson in the sulky—won the first heat going away."

Also from the *Washington Post,* this headline: "Antibiotic-Combination Drugs Used to Treat Colds Banned by FDA." Perhaps the next FDA regulation will ban all colds, and virologists will have to find a different line of work.

A manuscript contained this sentence: "A large mass of literature has accumulated on the cell walls of staphylococci." After the librarians have catalogued the staphylococci, they will have to start on the fish, according to this sentence from a manuscript: "The resulting disease has been described in detail in salmon."

A book review contained this sentence: "This book includes discussion of shock and renal failure in separate chapters."

The first paragraph of a news release issued by the American Lung Association said, "'Women seem to be smoking more but breathing less,' says Colin R. Woolf, M.D., Professor, Department of Medicine, University of Toronto. He presented evidence that women who smoke are likely to have pulmonary abnormalities and impaired lung function at the annual meeting of the American Lung Association." Even though the annual meeting was in the lovely city of Montreal, we hope that women who smoke stayed home.

And finally, some favorites from an email discussion list: A student wrote that she was seeking housing "for me, my two dogs and my rabbit that has a washer dryer." (We hadn't realized that rabbits do laundry.) A technician said she was looking for a "large or medium dog kennel for a researcher." (Hmmm, is office space that scarce?) And another list member wrote that "due to moving, internship salary, and a lack of need for it," she was highly motivated to sell the item that she was advertising. (Gee, if you don't need your internship salary, we know someone who would like it.)

THE TEN COMMANDMENTS OF GOOD WRITING

1. Each pronoun should agree with their antecedent.
2. Just between you and I, case is important.
3. A preposition is a poor word to end a sentence with. (Incidentally, did you hear about the streetwalker who violated a grammatical rule? She unwittingly approached a plainclothesman, and her proposition ended with a sentence.)
4. Verbs has to agree with their subject.
5. Don't use no double negatives.
6. Remember to never split an infinitive.
7. Avoid clichés like the plague.
8. Join clauses good, like a conjunction should.
9. Do not use hyperbole; not one writer in a million can use it effectively.
10. About sentence fragments.

Actually, the following story from one of us (R.A.D.) may change some minds about the use of double negatives: During the last presidential election, I visited my old hometown, which is in the middle of a huge cornfield in northern Illinois. Arriving after a lapse of some years, I was pleased to find that I could still understand the natives. In fact, I was a bit shocked to find that their language was truly expressive even though they were blissfully unaware of the rule against double negatives. One evening at the local gathering place, appropriately named the Farmer's Tavern, I orated at the man on the next bar stool about the relative demerits of the two presidential candidates. His lack of interest was then communicated in the clear statement; "Ain't nobody here knows nothin' about politics." While I was savoring this triple negative, a morose gent at the end of the bar looked soulfully into his beer and proclaimed: "Ain't nobody here knows nothin' about nothin' nohow." Strangely, this quintuple negative provided the best description I have ever heard of my hometown.

METAPHORICALLY SPEAKING

Although metaphors are not covered by the above rules, we suggest that you largely avoid similes and metaphors. Use them rarely in scientific writing. If you use them, use them carefully. We have all seen mixed metaphors and noted how comprehension gets mixed along with the metaphor. (Figure this one out: A virgin forest is a place where the hand of man has never set foot.) A rarity along this line is the "self-canceling metaphor." A favorite was ingeniously concocted by the eminent microbiologist L. Joe Berry. After one of his suggestions had been quickly negated by a committee vote, Joe said, "Boy, I got shot down in flames before I ever got off the ground."

Watch for hackneyed expressions. These are usually similes or metaphors (for example, timid as a mouse). Interesting and picturesque writing results from the use of fresh similes and metaphors; dull writing results from the use of stale ones.

Some words have become hackneyed, usually by being hopelessly locked to some other word. One example is the word "leap"; a "leap" is insignificant unless it is a "quantum leap." Another example is the verb "wreak." One can "wreak havoc" but nothing else seems to get wreaked these days. Since the dictionary says that "wreak" means "to bring about," one should be able to "wreak a weak pain for a week." To wreak a wry smile, try saying "I've got a weak back." When someone asks when you got it, you respond, "Oh, about a week back." (At the local deli, we call this tongue in cheek on wry.) That person may then respond, "Wow. That boggles the mind." You can then cleverly ask what else gets boggled these days.

MISUSE OF WORDS

Also watch for self-canceling or redundant words. Recently someone was described as being a "well-seasoned novice." A newspaper article referred to "young juveniles." A sign in a stamp and coin dealer's shop read "authentic replicas." If there is any expression that is dumber than "7 a.m. in the morning," it is "viable alternative." (If an alternative is not viable, it is not an alternative.)

Certain words are wrongly used thousands of times in scientific writing. Some of the worst offenders are the following:

amount. Use this word when you refer to a mass or aggregate. Use "number" when individual entities are involved. "An amount of cash" is all right. "An amount of coins" is wrong.

and/or. This is a slipshod construction used by thousands of authors but accepted by few experienced editors. Bernstein (1965) said, "Whatever its uses in legal or commercial English, this combination is a visual and mental monstrosity that should be avoided in other kinds of writing."

it. This common, useful pronoun can cause a problem if an antecedent is not clear, as in the sign that read: "Free information about VD. To get it, call 555-7000."

like. Often used incorrectly as a conjunction. Should be used only as a preposition. When a conjunction is needed, substitute "as." Like I just said, this sentence should have started with "As."

only. Many sentences are only partially comprehensible because the word *only* is positioned correctly in the sentence only some of the time. Consider this sentence: "I hit him in the eye yesterday." The word *only* can be added at the start of the sentence, at the end of the sentence, or between any two words within the sentence, but look at the differences in meaning that result.

quite. Next time you notice this word in one of your manuscripts, delete it and read the sentence again. You will notice that quite is quite unnecessary.

varying. The word "varying" means "changing." Often used erroneously when "various" is meant. "Various concentrations" are defined concentrations that do not vary.

which. The word "which" is properly used in a "nonrestrictive" sense, to introduce a clause that is not essential to the rest of the sentence; "that" introduces an essential clause. Examine these two sentences: "CetB mutants, *which* are tolerant to colicin E2, also have an altered. . . ." "CetB mutants *that* are tolerant to colicin E2 also have an altered. . . ." Note the substantial difference in meaning. The first sentence indicates that *all* CetB mutants are tolerant to colicin; the second sentence indicates that only some of the CetB mutants are tolerant to it.

while. When a time relationship exists, "while" is correct; otherwise, "whereas" would be a better choice. "Nero fiddled while Rome burned" is fine. "Nero fiddled while we wrote a book on scientific writing" is not.

Those of us who have struggled to make ourselves understood in a foreign language might especially appreciate this story: A graduate student had recently arrived in the United States from one of the more remote countries of the world. He had a massive English vocabulary, developed by many years of assiduous study. Unfortunately, he had had few opportunities to speak the language. Soon after his arrival, the dean of the school invited a number of the students and faculty to an afternoon tea. Some of the faculty members soon engaged the new foreign student in conversation. One of the first questions asked was, "Are you married?" The student said, "Oh, yes, I am most entrancingly married to one of the most exquisite belles of my country, who will soon be arriving here in the United States, ending our temporary bifurcation." The faculty members exchanged questioning glances—then came the next question: "Do you have children?" The student answered "No." After some thought, the student decided this answer needed some amplification, so he said, "You see, my wife is inconceivable." At this, his questioners could not hide their smiles, so the student, realizing he had committed a faux pas, decided to try again. He said, "Perhaps I should have said that my wife is impregnable." When this comment was greeted with open laughter, the student decided to try one more time: "I guess I should have said my wife is unbearable."

All seriousness aside, is there something about the use (rather than abuse) of English in scientific writing that merits special comment? The following is a tense answer.

TENSE IN SCIENTIFIC WRITING

One special convention of writing scientific papers is very tricky. It has to do with tense, and it is important because proper usage derives from scientific ethics.

When a scientific paper has been validly published in a primary journal, it thereby becomes knowledge. Whenever you state previously published findings, ethics requires you to treat the work with respect. You do this by using the *present* tense. It is correct to say "Streptomycin inhibits the growth of *M. tuberculosis* (13)." Whenever you state previously published findings, you should use the present tense; you are referring to established knowledge. You would do this just as you would say "The Earth is round." (If previously published results have been proven false by later experiments, the use of past rather than present tense would be appropriate.)

Your own present work must be referred to in the *past* tense. Your work is not presumed to be established knowledge until *after* it has been published. If you determined that the optimal growth temperature for *Streptomyces everycolor* was 37°C, you should say "*S. everycolor* grew best at 37°C." If you are citing previous work, possibly your own, it is then correct to say "*S. everycolor* grows best at 37°C."

In the typical paper, you will normally go back and forth between the past and present tenses. Most of the abstract should be in the past tense, because you are referring to your own present results. Likewise, the materials and methods and the results sections should be in the past tense, as you describe what you did and what you found. On the other hand, much of the introduction and much of the discussion should be in the present tense, because these sections often emphasize previously established knowledge.

Suppose that your research concerned the effect of streptomycin on *Streptomyces everycolor.* The tense would vary somewhat as follows.

In the abstract you would write, "The effect of streptomycin on *S. everycolor* grown in various media *was* tested. Growth of *S. everycolor,* measured in terms of optical density, *was* inhibited in all media tested. Inhibition *was* most pronounced at high pH levels."

In the introduction, typical sentences might be, "Streptomycin *is* an antibiotic produced by *Streptomyces griseus* (13). This antibiotic *inhibits* the growth of certain other strains of *Streptomyces* (7, 14, 17). The effect of streptomycin on *S. everycolor is* reported in this paper."

In the materials and methods section you would write, "The effect of streptomycin *was* tested against *S. everycolor* grown on Trypticase soy agar (BBL) and several other media (Table 1). Various growth temperatures and pH levels *were* employed. Growth *was* measured in terms of optical density (Klett units)."

In the results you would write, "Growth of *S. everycolor was* inhibited by streptomycin at all concentrations tested (Table 2) and at all pH levels (Table 3). Maximum inhibition *occurred* at pH 8.2; inhibition *was* slight below pH 7."

In the discussion you might write, "*S. everycolor was* most susceptible to streptomycin at pH 8.2, whereas *S. nocolor is* most susceptible at pH 7.6 (13). Various other *Streptomyces* species *are* most susceptible to streptomycin at even lower pH levels (6, 9, 17)."

In short, you should normally use the present tense when you refer to previously published work, and you should use the past tense when referring to your present results.

The main exceptions to this rule are in the areas of attribution and presentation. It is correct to say, "Smith (9) *showed* that streptomycin inhibits *S. nocolor.*" It is also correct to say, "Table 4 *shows* that streptomycin inhibited *S. everycolor* at all pH levels." Another exception is that the results of calculations and statistical analysis should be in the present tense, even though statements about the objects to which they refer are in the past tense; for example, "These values *are* significantly greater than those of the females of the same age, indicating that the males *grew* more rapidly." Still another exception is a general statement or known truth. Simply put, you could say, "Water *was added* and the towels *became* damp, which proves again that water *is* wet." More commonly, you will need to use this kind of tense variation: "Significant amounts of type IV procollagen *were* isolated. These results *indicate* that type IV procollagen *is* a major constituent of the Schwann cell ECM."

ACTIVE VERSUS PASSIVE VOICE

Let us now talk about *voice*. In any type of writing, the active voice is usually more precise and less wordy than is the passive voice. (This is not always true; if it were, we would have an Eleventh Commandment: "The passive voice should never be used.")

As noted in Chapter 11, passive voice sometimes functions well in the methods section. Elsewhere in a scientific paper, however, it rarely should be chosen.

Why, then, do scientists use so much passive voice? Perhaps this bad habit results from the erroneous idea that it is somehow impolite to use first-person pronouns. Because of this idea, the scientist commonly uses verbose (and imprecise) statements such as, "It was found that" in preference to the short, unambiguous "I found."

Young scientists should renounce the false modesty of their predecessors. Do not be afraid to name the agent of the action in a sentence, even when it is "I" or "we." Once you get into the habit of saying "I found," you will also find that you tend to write "*S. aureus* produced lactate" rather than "Lactate was produced by *S. aureus*." (Note that the "active" statement is in three words; the passive requires five.)

You can avoid the passive voice by saying, "The authors found" instead of "it was found." Compared with the simple "we," however, "the authors" is pretentious, verbose, and imprecise (which authors?).

EUPHEMISMS

In scientific writing, euphemistic words and phrases normally should be avoided. The harsh reality of dying is not improved by substituting "passed away." Laboratory animals are not "sacrificed," as though scientists engaged in arcane religious exercises. They are killed and that's that. The *CBE Style Manual* (CBE Style Manual Committee 1983) cited a beautiful example of this type of euphemism: "Some in the population suffered mortal consequences from the lead in the flour." The *Manual* then corrects this sentence, adding considerable clarity as well as eliminating the euphemism: "Some people died as a result of eating bread made from the lead-contaminated flour." An instructor gave graduate students the "mortal consequences" sentence as a test question in scientific writing. Most were simply unable to say "died." On the other hand, there were some inventive answers. They included "Get the lead out" and "Some were dead from the lead in the bread."

SINGULARS AND PLURALS

If you use first-person pronouns, use both the singular and the plural forms as needed. Do not use the "editorial we" in place of "I." The use of "we" by a single author is outrageously pedantic.

A frequent error in scientific papers is the use of plural forms of verbs when the singular forms would be correct. For example, you should say, "10 g *was* added," not "10 g *were* added." This is because a *single* quantity was added. Only if the 10 g were added 1 g at a time would it be correct to say, "10 g were added."

The singular-plural problem also applies to nouns. The problem is severe in scientific writing, especially in biology, because so many of the words are, or are derived from, Latin or Greek. Commonly these words retain their Latin or Greek plurals; at least they do when used by careful writers.

Many of these words (for example: *data, media*) have entered popular speech, where the Latin "a" plural ending is rarely recognized as a plural. Most people habitually use "data is" constructions and probably have never used the real singular, *datum*. In *The Careful Writer,* Bernstein (1965) objected to this usage, terming it "a common solecism." Today, although debate on the subject persists, both the plural and the singular usages are commonly acceptable, at least in informal contexts. For instance, *Merriam-Webster's Collegiate Dictionary* (11th edition) gives "the data is plentiful" as an example of accepted usage. Whether to use *data* as a plural or a singular can depend on whether you are referring to a group of individual pieces of data or the data as a single mass. Often, in deciding whether to use *data* with a plural verb or a singular one, the best approach is to follow your discipline's conventions in this regard and the predominant usage in your target journal.

(www.phdcomics.com)

This "plural" problem was commented upon by Sir Ashley Miles, eminent microbiologist and scholar, in a letter to the editor of *ASM News* (*44*:600, 1978):

> *A Memoranda on Bacterial Motility.* The motility of a bacteria is a phenomena receiving much attention, especially in relation to the structure of a flagella and the effect on it of an antisera. No single explanatory data is available; no one criteria of proof is recognized; even the best media to use is unknown; and no survey of the various levels of scientific approach indicates any one strata, or the several stratae, from which answers may emerge. Flagellae are just as puzzling as the bacteriae which carry them.

NOUN PROBLEMS

Another frequent problem in scientific writing is the verbosity that results from use of abstract nouns. This malady is corrected by turning the nouns into verbs. "Examination of the patients was carried out" should be changed to the more direct "I examined the patients"; "separation of the compounds was accomplished" can be changed to "the compounds were separated"; "transformation of the equations was achieved" can be changed to "the equations were transformed."

An additional problem with nouns results from using them as adjectives. Normally, there is no problem with such usage, but you should watch for special difficulties. For example, there is no problem in understanding "liver disease" (even though the adjective "hepatic" could be substituted for the noun "liver"). The problem aspect is illustrated by the following sentences from an autobiography: "When I was 10 years old, my parents sent me to a child psychiatrist. I went for a year and a half. The kid didn't help me at all." There once was an ad (in the *New York Times,* of all places) with the headline "Good News for Home Sewers." It could have been an ad for a drain-cleaning compound or for needle and thread.

The problem gets still worse when clusters of nouns are used as adjectives, especially when a real adjective gets into the brew. "Tissue culture response" is awkward; "infected tissue culture response" may be baffling. (Just what is infected?) Baffling too are these gems from job ads: "newborn hospital photographer" and "portable toilet route driver." Sometimes you can resolve the confusion by inserting a hyphen to show which words function together as an adjective. Consider, for example, the headline "Technology can help drought hit farmers." (How odd to use technology to aid drought in hitting farmers!) Inserting a well-placed hyphen yields a much more reasonable headline: "Technology can help drought-hit farmers."

NUMBERS

Preferred usage regarding numbers varies among style manuals and among journals. *The Chicago Manual of Style* (2010) favors spelling out "whole numbers from one through one hundred" and using numerals for other numbers. However, it notes that many publications, for example in science, spell out only single-digit numbers. In the "revised or modern scientific number style" (Style Manual Subcommittee, Council of Science Editors 2014), single-digit whole numbers, with few exceptions, appear as numerals too.

If style for numbers is not specified otherwise, here are some widely acceptable guidelines to follow: One-digit numbers should be spelled out; numbers of two or more digits should be expressed as numerals. You would write "three experiments" or "13 experiments." Now the exception: With standard units of measure, always use numerals. You would write "3 ml" or "13 ml." The only exception to the exception is that you should not start a sentence with a numeral. You should either reword the sentence or spell out both the number and the unit of measurement. For example, your sentence could start "Reagent A (3 ml) was added" or it could start "Three milliliters of reagent A was added." Actually, there is still another exception, although it comes up rarely. In a sentence containing a series of numbers, at least one of which is of more than one digit, all of the numbers should be expressed as numerals. (Example: "I gave water to 3 scientists, milk to 6 scientists, and beer to 11 scientists.")

ODDS AND ENDS

Apropos of nothing, one might mention that English is a strange language. Isn't it curious that the past tense of "have" ("had") is converted to the past participle simply by repetition: He *had had* a serious illness. Strangely, it is possible to string together 11 "hads" in a row in a grammatically correct sentence. If one were to describe a teacher's reaction to papers turned in by students John and Jim, one could say: John, where Jim had had "had," had had "had had"; "had had" had had an unusual effect on the teacher. That peculiar word "that" can also be strung together, as in this sentence: He said, in speaking of the word "that," that that "that" that that student referred to was not that "that" that that other student referred to.

The "hads" and the "thats" in a row show the power of punctuation. As a further illustration, here is a little grammatical game that you might want to try on your friends. Hand a slip of paper to each person in the group and ask the members of the group to provide any punctuation necessary to the following seven-word sentence: "Woman without her man is a savage." Some members will quickly respond that the sentence needs no punctuation, and they are

(By permission of Johnny Hart and Creators Syndicate, Inc.)

correct. A few pedants in the group will place balancing commas around the prepositional phrase: "Woman, without her man, is a savage." Grammatically, this is also correct. Other group members, however, will place a dash after "woman" and a comma after "her." Then we have "Woman—without her, man is a savage." It too is a correct response.

Seriously, we should all come to understand that sexism in language can have "savage" results. Scientific writing that promotes stereotypes is not scientific. Good guides have been published to show us how to avoid use of sexist and other biased language (Schwartz et al. 1995; Maggio 1997). Some style manuals also provide guidance regarding this issue.

Let us end where we started by again emphasizing the importance of syntax. Whenever comprehension goes out the window, faulty syntax is usually responsible. Sometimes, faulty syntax is simply funny and comprehension is not lost, as in these two items, culled from want ads: "For sale, fine German Shepherd dog, obedient, well trained, will eat anything, very fond of children." "For sale, fine grand piano, by a lady, with three legs."

But look at this sentence, which is similar to thousands that have appeared in the scientific literature: "Thymic humoral factor (THF) is a single heat-stable polypeptide isolated from calf thymus composed of 31 amino acids with molecular weight of 3,200." The prepositional phrase "with molecular weight of 3,200" would logically modify the preceding noun "acids," meaning that the amino acids had a molecular weight of 3,200. Less logically, perhaps the calf thymus had a molecular weight of 3,200. Least logical of all (because of their distance apart in the sentence) would be for the THF to have a molecular weight of 3,200—but, indeed, that was what the author was trying to tell us.

If you have any interest whatsoever in learning to use English more effectively, you should read Strunk and White's (2000) *The Elements of Style*. The "elements" are given briefly (in less than 100 pages!) and clearly. Anyone writing anything should read and use this famous little book. (You can read an early edition, by Strunk alone, at www.bartleby.com/141/.) After you have mastered Strunk and White, proceed immediately to Fowler (1965). Do not pass go; do not collect $200. Of course, if you really do want to get a Monopoly on good scientific English, buy that superbly quintessential book, *Scientific English* (Day and Sakaduski 2011).

CHAPTER 31

Avoiding Jargon

Clutter is the disease of American writing. We are a society strangling in unnecessary words, circular constructions, pompous frills and meaningless jargon.
—William Zinsser

DEFINITION OF JARGON

According to *Merriam-Webster's Collegiate Dictionary* (11th edition), definitions of jargon include the following: (1) "confused unintelligible language"; (2) "the technical terminology or characteristic idiom of a special activity or group"; (3) "obscure and often pretentious language marked by circumlocutions and long words."

The first and third types of jargon should be avoided. The second type ("technical terminology") is much more difficult to avoid in scientific writing, and it may be used if readers already are familiar with it or if you have defined or explained it. If you are writing for a technically trained audience, only the unusual technical terms need explanation.

MUMBLESPEAK AND OTHER SINS

The most common type of verbosity that afflicts authors is jargon. This syndrome is characterized, in extreme cases, by the total omission of one-syllable words. Writers with this affliction never *use* anything—they *utilize*. They never *do*—they *perform*. They never *start*—they *initiate*. They never *end*—they *finalize* (or *terminate*). They never *make*—they *fabricate*. They use *initial* for *first*, *ultimate* for *last, prior to* for *before, subsequent to* for *after, militate against* for *prohibit, sufficient* for *enough,* and *a plethora* for *too much.* An occasional author will slip and use the word *drug,* but most will salivate like Pavlov's dogs in anticipation of using *chemotherapeutic agent.* (We do hope that the name Pavlov rings a bell.) Who would use the three-letter word *now* instead of the elegant expression *at this point in time?*

Stuart Chase (1954) told the story of the plumber who wrote to the Bureau of Standards saying he had found that hydrochloric acid is good for cleaning out clogged drains. The bureau wrote back, "The efficacy of hydrochloric acid is indisputable, but the chlorine residue is incompatible with metallic permanence." The plumber replied that he was glad the bureau agreed. The bureau tried again, writing, "We cannot assume responsibility for the production of toxic and noxious residues with hydrochloric acid, and suggest that you use an alternate procedure." The plumber again said that he was glad that the bureau agreed with him. Finally, the bureau wrote to the plumber, "Don't use hydrochloric acid; it eats hell out of the pipes."

Should we liken the scientist to a plumber, or is the scientist perhaps more exalted? With that doctor of philosophy degree, should the scientist know some philosophy? We agree with John W. Gardner, who said, "The society which scorns excellence in plumbing because plumbing is a humble activity and tolerates shoddiness in philosophy because it is an exalted activity will have neither good plumbing nor good philosophy. Neither its pipes nor its theories will hold water" (*Science News,* p. 137, March 2, 1974).

We like the way that Aaronson (1977) put it: "But too often the jargon of scientific specialists is like political rhetoric and bureaucratic mumble-speak: ugly-sounding, difficult to understand, and clumsy. Those who use it often do so because they prefer pretentious, abstract words to simple, concrete ones."

The trouble with jargon is that it is a special language, the meaning of which is known only to a specialized "in" group. Science should be universal, and therefore every scientific paper should be written in a universal language.

Of course, you will have to use specialized terminology on occasion. If such terminology is readily understandable to practitioners and students in the field, there is no problem. If the terminology is *not* recognizable to any portion of your potential audience, you should (1) use simpler terminology or (2) carefully define the esoteric terms (jargon) that you are using. In short, you should not write for the half-dozen or so people who are doing exactly your kind of work. You should write for the hundreds of people whose work is only slightly related to yours but who may want or need to know something about your work.

MOTTOES TO LIVE BY

Here are a few important concepts that all readers of this book should master. They are, however, expressed in typical scientific jargon. With a little effort you can probably translate these sentences into simple English:

1. As a case in point, other authorities have proposed that slumbering canines are best left in a recumbent position.
2. An incredibly insatiable desire to understand that which was going on led to the demise of this particular *Felis catus.*
3. There is a large body of experimental evidence which clearly indicates that members of the genus *Mus* tend to engage in recreational activity while the feline is remote from the locale.
4. From time immemorial, it has been known that the ingestion of an "apple" (that is, the pome fruit of any tree of the genus *Malus,* said fruit being usually round in shape and red, yellow, or greenish in color) on a diurnal basis will with absolute certainty keep a primary member of the health care establishment absent from one's local environment.
5. Even with the most sophisticated experimental protocol, it is exceedingly unlikely that the capacity to perform novel feats of legerdemain can be instilled in a superannuated canine.
6. A sedimentary conglomerate in motion down a declivity gains no addition of mossy material.
7. The resultant experimental data indicate that there is no utility in belaboring a deceased equine.

If you had trouble with any of the above, here are the jargon-free translations:

1. Let sleeping dogs lie.
2. Curiosity killed the cat.
3. When the cat's away, the mice will play.
4. An apple a day keeps the doctor away.
5. You can't teach old dogs new tricks.
6. A rolling stone gathers no moss.
7. Don't beat a dead horse.

BUREAUCRATESE

Regrettably, too much scientific writing fits the first and third definitions of jargon. All too often, scientists write like the legendary Henry B. Quill, the bureaucrat described by Meyer (1977): "Quill had mastered the mother tongue of government. He smothered his verbs, camouflaged his subjects and hid everything in an undergrowth of modifiers. He braided, beaded and fringed, giving elaborate expression to negligible thoughts, weasling [*sic*], hedging and announcing the obvious. He spread generality like flood waters in a long, low valley. He sprinkled everything with aspects, feasibilities, alternatives, effectuations, analyzations, maximizations, implementations, contraindications and appurtenances. At his best, complete immobility set in, lasting sometimes for dozens of pages."

Some jargon, or bureaucratese, consists of clear, simple words but contains so many words that the meaning is not readily evident. Examine the following, an important federal regulation (*Code of Federal Regulations,* Title 36, Paragraph 50.10) designed to protect trees from injury; this notice was posted in National Capital Park and Planning Commission recreation areas in the Washington area:

TREES, SHRUBS, PLANTS, GRASS AND OTHER VEGETATION

> (a) General Injury. No person shall prune, cut, carry away, pull up, dig, fell, bore, chop, saw, chip, pick, move, sever, climb, molest, take, break, deface, destroy, set fire to, burn, scorch, carve, paint, mark, or in any manner interfere with, tamper, mutilate, misuse, disturb or damage any tree, shrub, plant, grass, flower, or part thereof, nor shall any person permit any chemical, whether solid, fluid or gaseous to seep, drip, drain or be emptied, sprayed, dusted or injected upon, about or into any tree, shrub, plant, grass, flower or part thereof except when specifically authorized by competent authority; nor shall any person build fires or station or use any tar kettle, heater, road roller or other engine within an area covered by this part in such a manner that the vapor, fumes or heat therefrom may injure any tree or other vegetation.
>
> (TRANSLATION: Don't mess with growing things.)

Jargon does not necessarily involve the use of specialized words. Faced with a choice of two words, the jargonist always selects the longer one. The jargonist really gets his jollies, however, by turning short, simple statements into long strings of words. And, usually, the longer word or the longer series of words is not as clear as the simpler expression. We challenge anyone to show how "at this point in time" means, in its cumbersome way, more than the simple word "now." The concept denoted by "if" is not improved by substituting the pompous expression "in the event that."

SPECIAL CASES

Perhaps the worst offender is the word "case." There is no problem with a case of canned goods or even a case of flu. However, 99 percent of the uses of "case" are jargon. In case you think that 99 percent is too high, make your own study. Even if this percentage is too high, a good case could be made for the fact that "case" is used in too many cases. Better and shorter usage should be substituted: "in this case" means "here"; "in most cases" means "usually"; "in all cases" means "always"; "in no cases" means "never." (We also have issues with "issues." And we wish to highlight "highlight." In formal writing, limit use of these words to their specific meanings.)

Still another word that causes trouble (in some cases) is "about," not because it is used but because it is avoided. As pointed out by Weiss (1982), writers seem unwilling to use the clear, plain "about" and instead use wordier and less-clear substitutes such as the following:

approximately	pursuant to
in connection with	re
in reference to	reference
in relation to	regarding
in the matter of	relating to the subject matter of
in the range of	relative to
in the vicinity of	respecting
more or less	within the ballpark of
on the order of	with regard to
on the subject of	with respect to

Appendix 2 contains some "Words and Expressions to Avoid." A similar list well worth consulting was published by O'Connor and Woodford (1975). It is not necessarily improper to use any of these words or expressions *occasionally;* if you use them repeatedly, however, you are writing in jargon and your readers are suffering.

Perhaps the most common way of creating a new word is the jargonist's habit of turning nouns into verbs. One example is use of the word "interface" to mean "communicate"; the only time people can interface is when they kiss. And a classic example appeared in a manuscript that read: "One risks exposure when swimming in ponds or streams near which cattle have been pasturized." The copy editor, knowing that there is no such word as "pasturized," changed it to "pasteurized." (There may be nothing wrong with that. If you can pasteurize milk, presumably you can pasteurize the original container.)

In their own pastures, scientists are, of course, very expert, but they often succumb to pedantic, jargonistic, and useless expressions, telling the reader more than the reader wants or needs to know. As the English novelist George Eliot said: "Blessed is the man who, having nothing to say, abstains from giving us wordy evidence of this fact."

If you must show off your marvelous vocabulary, make sure you use the right words. Consider the story that Lederer (1987) told about NASA scientist Wernher von Braun. "After one of his talks, von Braun found himself clinking cocktail glasses with an adoring woman from the audience.

" 'Dr. von Braun,' the woman gushed, 'I just loved your speech, and I found it of absolutely infinitesimal value!'

" 'Well then,' von Braun gulped, 'I guess I'll have it published posthumously.'

" 'Oh yes!' the woman came right back. 'And the sooner the better.' "

Or consider the two adventuresome hot-air balloonists who, slowly descending after a long trip on a cloudy day, looked at the terrain below and had not the faintest idea where they were. It so happens that they were drifting over the grounds of one of our more famous scientific research institutes. When the balloonists saw a man walking along the side of a road, one called out, "Hey, mister, where are we?" The man looked up, took in the situation, and, after a few moments of reflection, said, "You're in a hot-air balloon." One balloonist turned to the other and said, "I'll bet that man is a scientist." The other balloonist asked, "What makes you think so?" To which the first replied, "His answer is perfectly accurate—and totally useless."

CHAPTER 32

How and When to Use Abbreviations

Authors who use abbreviations extravagantly need to be restrained.
—Maeve O'Connor

GENERAL PRINCIPLES

Many experienced editors loathe abbreviations. Some editors would prefer that they not be used at all, except for standard units of measurement and their Système International (SI) prefixes, for which all scientific journals allow abbreviations. Many journals also allow, without definition, such standard abbreviations as etc., et al., i.e., and e.g. (The abbreviations i.e. and e.g. are often misused; properly used, i.e. means "that is," whereas e.g. means "for example." Because these abbreviations are so often misused or misinterpreted, we favor avoiding them.) In your own writing, you would be wise to keep abbreviations to a minimum. The editor will look more kindly on your paper, and the readers of your paper will bless you forever. More preaching on this point should not be necessary because, by now, you yourself have no doubt come across undefined and indecipherable abbreviations in the literature. Just remember how annoyed you felt when you were faced with these conundrums, and join with us now in a vow never again to pollute the scientific literature with an undefined abbreviation.

The "how to" of using abbreviations is easy, because most journals use the same convention. When you plan to use an abbreviation, you introduce it by spelling out the word or term first, followed by the abbreviation within parentheses. The first sentence of the introduction of a paper might read: "Bacterial plasmids, as autonomously replicating deoxyribonucleic acid (DNA) molecules of modest size, are promising models for studying DNA replication and its control."

The "when to" of using abbreviations is much more difficult. Several general guidelines might help.

First, generally do not use an abbreviation in the title of an article. Very few journals allow abbreviations in titles, and their use is strongly discouraged by the indexing and abstracting services. If the abbreviation is not standard, the literature retrieval services will have a difficult or impossible problem. Even if the abbreviation is standard, indexing and other problems arise. One major problem is that accepted abbreviations have a habit of changing; today's abbreviations may be unrecognizable a few years from today. Comparison of certain abbreviations as listed in the various editions of the *Council of Biology Editors Style Manual* (which has now become *Scientific Style and Format: The CSE Manual for Authors, Editors, and Publishers*) emphasizes this point. Dramatic changes occur when the terminology itself changes. Students today could have trouble with the abbreviation DPN (which stands for "diphosphopyridine nucleotide"), because the name itself has changed to "nicotinamide adenine dinucleotide," the abbreviation for which is NAD.

Abbreviations should almost never be used in the abstract. Only if you use the same name, a long one, many times should you consider an abbreviation. If you use an abbreviation, you must define it at the first use in the abstract. Remember that the abstract will stand alone in whichever abstracting databases cover the journal in which your paper appears.

In the text itself, abbreviations may be used if a long word or phrase will appear repeatedly. They serve a purpose in reducing printing costs, by somewhat shortening the paper. More importantly, they aid the reader when they are used judiciously. Speaking of "importantly": We are reminded of a man whose children sometimes refer to him as "the FIP" (fairly important person). They know that he hasn't yet made it to VIP.

GOOD PRACTICE

It can be advisable, when writing the first draft of the manuscript, to spell out all terms and phrases. Then examine the manuscript for repetition of long words or phrases that might be candidates for abbreviation. Do not abbreviate a term or phrase that is used only once or twice in the paper. If the term or phrase is used with modest frequency—for example, between three and six times—and a standard abbreviation for it exists, introduce and use the abbreviation. (Some journals allow some standard abbreviations to be used without definition at first use.) If no standard abbreviation exists, do not manufacture one unless the term or phrase is used frequently or is very long and cumbersome.

Often you can avoid abbreviations by using the appropriate pronoun (it, they, them) if the antecedent is clear. Another possibility is to use a substitute expression such as "the inhibitor," "the substrate," "the drug," "the enzyme," or "the acid."

Usually, you should introduce your abbreviations one by one as they first occur in the text. Alternatively, you might consider a separate paragraph (headed "Abbreviations Used") in the introduction or in the materials and methods section. The latter system (required in some journals) is especially useful if the names of related reagents, such as a group of organic chemicals, are to be used in abbreviated form later in the paper. Another option, for example in review papers and grant proposals, can be to include a table listing and defining abbreviations. Such tables make definitions easy to find even if a piece is not being read from beginning to end. Also, if chapters of a book might be read individually or in different orders, consider defining abbreviations on first appearance in each chapter. The same principle holds for other lengthy pieces of writing, such as long proposals, that might well be read other than from start to finish.

UNITS OF MEASUREMENT

Units of measurement are abbreviated when used with numerical values. You would write "4 mg was added." (The same abbreviation is used for the singular and the plural.) When used without numerals, however, units of measurement are not abbreviated. You would write "Specific activity is expressed as micrograms of adenosine triphosphate incorporated per milligram of protein per hour."

Careless use of the diagonal can cause confusion. This problem arises frequently in stating concentrations. If you say that "4 mg/ml of sodium sulfide was added," what does this mean? Does it mean "per milliliter of sodium sulfide" (the literal translation) or can we safely assume that "per milliliter of reaction mixture" is meant? It is much clearer to write "4 mg of sodium sulfide was added per milliliter of medium."

SPECIAL PROBLEMS

A frequent problem with abbreviations concerns use of "a" or "an." Should you write "a M.S. degree" or "an M.S. degree"? Recall the old rule that you use "a" with words beginning with a consonant sound and "an" with words beginning with a vowel sound (for example, the letter "em"). Because in science we should use only common abbreviations, those not needing to be spelled out in the reader's mind, the proper choice of article should relate to the sound of the first letter of the abbreviation, not the sound of the first letter of the spelled-out term. Thus, although it is correct to write "a master of science degree," it is incorrect to write "a M.S. degree." Because the reader reads "M.S." as "em ess," the proper construction is "an M.S. degree."

In biology, it is customary to abbreviate generic names of organisms after first use. At first use, you would spell out "*Streptomyces griseus.*" In later usage, you can abbreviate the genus name but not the specific epithet: *S. griseus.* Suppose, however, that you are writing a paper that concerns species of both *Streptomyces* and *Staphylococcus.* You would then spell out the genus names repeatedly. Otherwise, readers might be confused as to whether a particular "*S.*" abbreviation referred to one genus or the other.

SI (SYSTÈME INTERNATIONAL) UNITS

Appendix 3 gives the abbreviations for the prefixes used with all Système International (SI) units. The SI units and symbols, and certain derived SI units, have become part of the language of science. This modern metric system should be mastered by all students of the sciences. *Scientific Style and Format* (Style Manual Subcommittee, Council of Science Editors 2014) is a good source for more complete information, as is Huth's (1987) *Medical Style and Format.* Briefly, SI units include three classes of units: base units, supplementary units, and derived units. The seven base units that form the foundations of SI are the meter (or metre), kilogram, second, ampere, kelvin, mole, and candela. In addition to these seven base units, there are two supplementary units for plane and solid angles: the radian and steradian, respectively. Derived units are expressed algebraically in terms of base units or supplementary units. For some of the derived SI units, special names and symbols exist.

OTHER ABBREVIATIONS

Some style manuals in the sciences (for example, Iverson and others 2007; Style Manual Subcommittee, Council of Science Editors 2014) list abbreviations that are standard in specific fields. (On a related note: Such manuals also serve as resources on accepted nomenclature.) Use these and other standard abbreviations when strongly warranted. Largely avoid others. Those that you use should be introduced as carefully as you would introduce royalty.

CHAPTER 33

Writing Clearly across Cultures and Media

When you write for an Internet venue . . . your words travel the globe in a flash. But . . . what you think you're saying isn't always what the far-flung reader understands.

—Steve Outing

READABLE WRITING

Earlier chapters have presented principles of writing readably: structuring sentences simply, using proper syntax, deleting needless words, condensing wordy phrases, using words accurately, using mainly active voice, avoiding strings of nouns, using verbs rather than nouns made from them, punctuating properly, using short and familiar words, minimizing use of abbreviations, and defining abbreviations.

Also for readability, generally avoid starting sentences with "It is" or "There is." For example, change "It is not necessary to remove this structure" to "This structure need not be removed" or (if appropriate) "You need not remove this structure." Likewise, condense "There is another method that is gaining acceptance" to "Another method is gaining acceptance."

In general, say what things are, not what they are not. If you mean that something is important, do not say that it is "not unimportant." If you mean that it is substantial, do not say "not insubstantial." Avoiding such double negatives makes writing more readable.

Many suggestions for making writing more readable also make it shorter. This brevity can especially help if you have a word limit or page limit, such as for a scientific paper or grant proposal.

(© ScienceCartoonsPlus.com)

Over the years, formulas have been devised to estimate the readability of documents. Microsoft Word can compute two such measures of readability, the Flesch Reading Ease score and the Flesch-Kincaid Grade Level score. (Doing so is an option under "When correcting spelling and grammar," in "Proofing.") Computing these scores, which are based on average number of words per sentence and average number of syllables per word, can help you estimate how readable your writing is (or how much progress you have made in making it more readable). These formulas do not, however, take into account all aspects of readability. Thus, they are imperfect measures. Indeed, a piece of writing could make no sense but still get an excellent readability score if it consisted of short words in short sentences.

CONSISTENCY IN WORDING

For clarity in scientific writing, keep using the same word for the same thing. Do not feel compelled to vary your vocabulary, as you might in a literary piece, to make your writing more interesting. Readers should be able to focus on the content. They should not need to wonder whether "the mice," "the animals," and "the rodents" are the same creatures, or whether "the conference," "the convention," and "the meeting" are the same event. Using consistent wording can help make your writing clear and cohesive.

Some words, however—those that are so vivid or unusual that they tend to be remembered—should not be used repeatedly in close succession. In this regard, one can think of "blue jeans words" and "purple plaid trousers words," in keeping with this analogy presented to American graduate students:

> If you wore blue jeans to the laboratory every day, probably no one would notice that fact. Similarly, if you repeatedly used words such as "experiment," "molecule," "increase," and "journal," probably no one would notice. However, if you wore purple plaid trousers to the laboratory today, people probably would notice if you also did so next week. Similarly, if you used the word "astonishing," "armamentarium," "compendium," or "conundrum" in one paragraph, people probably would notice if you also did so in the next.

Stay mainly with blue jeans words, and feel free to use them repeatedly. Use purple plaid trousers words rarely, if at all.

SERVING INTERNATIONAL READERS

Consistent wording can especially help make your reading clear to readers whose native language is not English. Here are some other things you can do to help serve this readership: Use words that have one meaning or a few meanings, not many, and largely avoid idioms. (For example, in revising material for this book, "a good deal easier" was changed to "much easier"; "watch your similes and metaphors" was changed to "largely avoid similes and metaphors";

"do not bear repeated use" was changed to "should not be used repeatedly"; and "there is no bar" was changed to "there is no barrier.") Use mainly simple verb forms, and write sentences that are simply structured and not extremely long. Retain optional words that can clarify the structure of a sentence. For instance, write "I believe that Professor Day knows much about grammar," not "I believe Professor Day knows much about grammar," the first part of which might be misread as meaning that you believe Professor Day. Avoid literary and cultural allusions, including sports references, that might be unfamiliar to people in other cultures.

Additional guidance appears in *The Elements of International English Style* (Weiss 2005). Although geared more to the business and technology communities than to scientists, this book can aid in doing scientific writing that is clear to readers regardless of native language. It also can aid in communicating through letters and email to international colleagues.

A FEW WORDS ON EMAIL STYLE

If you are in the sciences, much of your writing probably is email. Although email is rarely published, a little attention to crafting it can help it serve you better.

Begin with a meaningful subject line. Then, for readability, keep the paragraphs fairly short, and skip space between them. Indeed, if you want your message to be read, make it relatively brief. Lengthy discourses often are better provided as attachments.

If you are sending email to a large group, spare readers the list of addresses by using the Bcc feature. And speaking of groups: In responding to messages sent to email discussion groups, beware of inadvertently replying to the whole group when you mean to address only the sender. The other members of the group probably do not care about your family vacation.

Beware of trying to convey humor by email: What may appear funny in person with vocal inflections or gestures may come across as hostile or otherwise offensive. You have better things to do than explain that you were really trying to be amusing.

If something in an email message annoys you, take time to cool down. Do not fire off an angry reply in haste. And angry or otherwise, do not say anything that you would not want forwarded. Remember that, other than in secure contexts, email is not private. As one colleague put it: If you would not write it on a postcard, do not put it in email. Clearly, email is not the medium for complaining about your graduate advisor or department head.

Although email tends to be casual, suit the level of formality to the context. When sending email to potential employers, for example, word it carefully, check it for grammatical errors, and proofread it thoroughly. If you have been using a humorous email address, consider having a more formal one for professional communications.

And finally, include a concise, informative signature block in your professional email. In the signature block, provide at least your name, title, and affiliation. If customary in your field or at your institution, perhaps also include a courtesy title (such as "Dr.") or list advanced degrees. Other items to consider providing in a signature block include your phone number, social media links, and URL. Consider having different signature blocks to use in different circumstances or modifying your signature block to suit the situation. In any case, keep your signature block relatively short. Remember, this is your signature block, not your curriculum vitae.

WRITING FOR ONLINE READING

The scientific papers you publish are likely to appear online. In addition, many of us in the sciences prepare material intended specifically for reading on the web. In preparing such items, consider the following pointers (Gahran 2000, 2001): Keep the material short, or break it into fairly self-sufficient chunks of 500 words or less. Consider starting with a synopsis to orient readers. Break long paragraphs into two or more paragraphs. Use clear headings to help readers find what they are seeking. Word links clearly and concisely. Consider highlighting key words. For readability, use bulleted (or numbered) lists instead of lists within paragraphs. Consider offering a printer-friendly version, containing text but not images.

If you have a *blog*—which is short, by the way, for "weblog"—also consider the following advice. Keep each post relatively short: in general, no more than 250 words (the equivalent of one double-spaced page). Give each post a title that is brief and informative; if feasible, make the title lively. Write in a consistent style. (An informal, conversational style generally suits blogs well. However, still be careful about spelling and grammar, and remember to proofread.) Provide posts at relatively regular intervals.

Because material posted on the web is accessible worldwide, writing in an internationally understandable way can be especially important. Therefore, keep sentences relatively short and direct, avoid regional idioms, and remember to define terms that might be unfamiliar to readers in other parts of the world (Outing 2001). By following such advice, you can make your material on the web truly a world wide resource.

CHAPTER 34

How to Write Science in English as a Foreign Language

ESL authors [authors for whom English is a second language] can be more precise about language just because it is their second language.

—Mary Boylan

ENGLISH AS THE INTERNATIONAL LANGUAGE OF SCIENCE

English is currently the international language of science. By no means does this demand that every scientific paper be written in English. Papers on findings mainly of local, national, or regional interest (for instance, in agricultural science, social science, or medicine) generally are best published in the language of those likely to use the content. However, when findings should be accessible to fellow scientists throughout the world, papers generally should appear in English.

For huge numbers of scientists, both in predominantly English-speaking countries and elsewhere, English is a second (or third, fourth, or fifth) language. In addition to facing the usual challenges of writing and publishing a scientific paper, these scientists face the challenges of writing in a foreign language and, oftentimes, interacting with editors from another culture. Yet many scientists from around the world have successfully met these challenges. This chapter, which is primarily for readers who are non-native speakers of English, is intended to aid in doing so. The chapter also may be useful to native-English-speaking scientists who want to work as effectively as possible with colleagues or students for whom English is not a native language.

THE ESSENTIALS: CONTENT, ORGANIZATION, AND CLARITY

Editors of good English-language journals want to publish the best science in the world, and many are eager to include work from a wide range of countries. Therefore, they often are willing to devote extra effort to publishing papers by non-native speakers of English (Iverson 2002). For example, they sometimes supplement peer reviewers' comments with detailed guidance of their own, and they sometimes allot extra staff time to copyediting papers that have good content but problems in English-language expression.

Thus, for non-native as well as native-English-speaking scientists, the editor and the author are allies. Do not be intimidated if you are a non-native speaker of English. If your research is of high quality and wide interest, editors of good English-language journals will want to publish it. Of course, you will have to do your part.

Your part consists mainly of submitting an informative, well-organized, clearly written paper. Some non-native speakers worry that their English seems unpolished or clumsy. In fact, some focus so much on making the English beautiful that they neglect more basic aspects. Although good English is certainly desirable, you need not agonize over fine points of style. If your paper is informative, well organized, and clear, the editor and peer reviewers can soundly evaluate your research. And then if your paper is accepted, a copy editor at the journal can readily correct occasional problems with grammar or other aspects of expression.

However, if important information is missing, if a paper is poorly organized, or if wording is unclear, the editor and peer reviewers might not be able to understand the paper well enough to evaluate the research. Even if they wish to publish the research, much difficult work may be needed to make the paper publishable. If the journal lacks the resources for this extra work, it might not be able to publish the paper. Even if it has such resources, such major difficulties may delay the paper's publication.

A copy editor at the journal may *query* you (ask you questions) if items in your paper are unclear or if he or she is uncertain whether proposed changes would retain your intended meaning. In the past, when queries and answers to them were routinely conveyed by postal service, the query process could substantially delay publication of papers by international authors. Now that email is widely available, scientists almost everywhere can receive and answer queries quickly. If your paper has been accepted for publication, check regularly for email messages, and reply promptly. If you do not understand a query, ask for clarification. Also, do not assume that the copy editor is right because he or she is an expert in English. He or she might have misunderstood you, and you are responsible for ensuring that your published paper is accurate.

CULTURAL DIFFERENCES TO CONSIDER

Cultures differ in a variety of norms relating to communication. Awareness of such differences can aid in writing and publishing your paper.

When manuscripts arrive from non-native speakers of English, issues sometimes arise about the level of detail included. For example, in manuscripts by authors from some countries, the materials and methods section tends not to be as informative as the journal requires. Cultures differ in how much information people supply, both in everyday conversation and in professional communications. Notice the level of detail, and types of details, in papers published in the journal to which you will submit your paper. Then write your paper accordingly.

Directness of expression also differs among cultures. In some cultures, expression tends to be indirect; the speaker or writer circles around the main point before eventually stating it—or maybe just implies the main point. In many Western cultures, however, and in leading international journals, expression tends to be direct, with the writer stating the main point and then providing details. In a typical paragraph in such a journal, a sentence at the beginning, known as the topic sentence, states the main point, and the other sentences in the paragraph then support that point or present related information. Before writing a paper for an English-language international journal, see how paragraphs in the journal tend to be structured. Then try to use that structure.

Cultures also differ in attitudes toward time. Some cultures greatly value speed and promptness, whereas others favor an unhurried pace. Prominent international journals typically embody the former attitudes. Therefore, reply quickly to inquiries from the journal, and take care to meet the deadlines that the journal sets—for example, for revising a manuscript. If you cannot meet a deadline, inform the editor as soon as possible, so he or she can plan accordingly.

Of concern to many editors is the fact that cultures also differ in attitudes toward using material taken word-for-word from other people's writing. In English-language scientific papers for international journals, authors are required to use their own wording for the vast majority of what they say and to clearly designate any wording taken from elsewhere. Thus, although authors may look at published papers to find words or phrases to use, they are not allowed to include entire passages from published work unless the passages are put in quotation marks (or, if long, indented) and the sources are cited. Otherwise, the author is considered guilty of *plagiarism*. A tutorial helpful in learning to recognize and avoid plagiarism appears at www.indiana.edu/~istd.

As noted in Chapter 5, steps for avoiding inadvertent plagiarism include clearly indicating in your notes the source of any material from others' work that you copy or download. If you inadvertently include in your paper a sentence or paragraph from elsewhere, a reviewer or copy editor might notice the difference in style and, to your embarrassment, ask whether the wording is your own. Woe to you if the passage happens to be by one of the peer reviewers!

SOME COMMON LANGUAGE CHALLENGES

In writing scientific papers, non-native speakers of English often face challenges relating to particular aspects of the English language—especially verb tenses, prepositions, and articles. With care, authors can minimize errors in these regards. Then, if the manuscript is clearly written, a copy editor at the journal can correct remaining errors.

Verb tenses, which differ among languages, often pose difficulty. As discussed in Chapter 30, the methods and results sections of a scientific paper should normally be written entirely, or almost entirely, in past tense. The introduction and discussion typically include a variety of tenses, depending on whether, for example, previously established knowledge is being presented (present tense) or the research reported in the present paper is being summarized (past tense). As well as following the general advice in this book, look at the use of verb tenses in the journal to which you are submitting your manuscript.

Deciding which preposition to use can be difficult, even sometimes for native speakers of English. Keeping a list of prepositional phrases commonly used in your field can help. So can consulting textbooks and websites intended to guide non-native speakers of English.

Likewise, proper use of articles ("a," "an," and "the") can be very difficult, especially for writers whose native languages do not contain articles. Here, too, it can help to consult textbooks and websites for users of English as a foreign language and to use published papers as examples. Other sources of guidance on various aspects of English include the book *Scientific English* (Day and Sakaduski 2011).

Other often-challenging aspects of English include plurals, mass nouns, capitalization, and sentence length and structure. Some authors from native languages without plural forms tend to forget to add an "s" to make English nouns plural. And some non-native speakers tend mistakenly to add an "s" to mass nouns (such as "information" and "research"). Native users of languages that do not have capital letters, or that follow different capitalization rules from those for English, sometimes neglect to capitalize English words when needed or capitalize excessively. Authors whose native languages tend to have very long sentences sometimes write sentences that should be several sentences in English. And sometimes non-native speakers use English words but retain the sentence structure of their native language, with awkward results. (A peer reviewer told an international colleague of ours that her sentences resembled those of the character Yoda!) If any of these aspects of English tend to pose difficulty for you, perhaps pay special attention to them when you revise your writing.

Native writers of some languages tend to have difficulty with spacing in English-language text. For example, sometimes they neglect to skip a space after the period at the end of a sentence, or they insert a space between an opening parenthesis and the word that follows, or they make many spacing errors in bibliographic references. If you tend to have this difficulty, check your manuscript carefully for proper spacing.

MORE STRATEGIES FOR ENGLISH-LANGUAGE WRITING

While teaching scientific writing overseas, an American instructor noticed that papers by one scientist in the class seemed almost as if they had been written by a native speaker of English. When the instructor commented on this fact, the scientist described his strategy: He carefully read several papers on his research topic in leading English-language journals and then, for each section (introduction, methods, etc.), listed words and phrases commonly used; when writing his papers, he consulted these lists. This strategy also can aid other non-native speakers of English. Likewise, keeping and consulting a list of revisions that copy editors or others have made in one's writing can help in polishing one's English.

Write simply overall. Do not try to impress readers with vocabulary words you have learned for the Test of English as a Foreign Language (TOEFL). Do not try to display your ability to write long, complex sentences in English. Do not try to exhibit your mastery of passive voice. Remember, the goal of a scientific paper is to communicate the science, not to impress readers with your English level. Many readers of your paper may be non-native speakers who know much less English than you do. Also, relatively simple writing makes a paper easier to understand even for native speakers of English, including editors and peer reviewers.

Draft your paper in English, if possible, rather than writing it in your native language and then translating it. Doing so can help your paper to read well in English. When you are drafting your paper, do not try to make the English perfect, as doing so can disrupt your flow of ideas. Rather, just try to express what you want to say. Then, once you have a draft, go back and, where necessary, improve the English.

If feasible, have someone with an especially strong command of English (and, ideally, knowledge of scientific writing and editing) review your paper before you submit it to a journal. (Indeed, if a paper seems to contain good science but is written in poor English, a journal may return the manuscript and suggest that it be edited by someone expert in English and then resubmitted.) If possible, the person providing feedback on your writing should be familiar with your field of science. Otherwise, although the person may correct grammar problems and other mechanical errors, he or she might not detect errors in scientific expression—and might inadvertently introduce errors (such as when one editor repeatedly changed the technical term "contracture" to "contraction"). Possible reviewers include colleagues at your institution or elsewhere who write well in English, professional editors at your institution, and teachers of scientific writing. Some professional English-language scientific-editing services exist. You also may be able to identify suitable editors through organizations such as the Council of Science Editors (www.CouncilScienceEditors.org), the European Association of Science Editors (www.ease.org.uk), and the Board of Editors in the Life Sciences (www.bels.org).

MORE RESOURCES

Many online resources can help non-native speakers write in English about science. One example is Academic Phrasebank (www.phrasebank.manchester.ac.uk), which lists phrases useful in various parts of a scientific paper. Another is Grammar Girl (www.quickanddirtytips.com/grammar-girl), which provides advice on grammar, punctuation, word choice, and related topics. Websites for users of English as a foreign language—for example, UsingEnglish.com (www.usingenglish.com)—also can be helpful.

You can find many such resources through the website of AuthorAID (www.authoraid.info), a project mainly to help researchers in developing countries write about and publish their work. The online resource library at this site includes links to many online resources. It also includes PowerPoint presentations, articles, and other materials on scientific writing and related subjects. In addition, the AuthorAID site contains a blog on communicating research. And through the site you can seek a mentor to advise you in your writing and related work. Although primarily for researchers in developing countries, the AuthorAID resources can also help researchers elsewhere with their writing. As a scientist from Japan said, "When it comes to scientific writing, every country is a developing country."

A goal of AuthorAID, like that of this book, is to increase researchers' knowledge, skill, and confidence regarding scientific writing and publication. If English is not your native language, do not feel discouraged. And when your paper is accepted by an international journal, consider celebrating twice: once in your native language and once in English.

PART VIII

Other Topics in Scientific Communication

CHAPTER 35

How to Write a Thesis

The average Ph.D. thesis is nothing but a transference of bones from one graveyard to another.

—J. Frank Dobie

PURPOSE OF THE THESIS

A Ph.D. thesis in the sciences is supposed to present the candidate's original research. Its purpose is to prove that the candidate can do and communicate such research. Therefore, a thesis should exhibit the same type of disciplined writing that is required in a journal publication. Unlike a scientific paper, a thesis may address more than one topic, and it may present more than one approach to some topics. The thesis may present all or most of the data obtained in the student's thesis-related research. Therefore, the thesis usually is longer and more involved than a scientific paper. But the concept that a thesis must be a bulky 200-page tome is wrong, dead wrong. Many 200-page theses contain only 50 pages of good science. The other 150 pages comprise turgid descriptions of insignificant details.

We have seen many Ph.D. theses, and we have assisted with the writing and organization of a good number of them. On the basis of this experience, we have concluded that there are almost no generally accepted rules for thesis preparation. Most types of scientific writing are highly structured. Thesis writing is not. The "right" way to write a thesis varies widely from institution to institution and even from professor to professor within the same department of the same institution.

Reid (1978) is one of many over the years who have suggested that the traditional thesis no longer serves a purpose. In Reid's words, "Requirements that

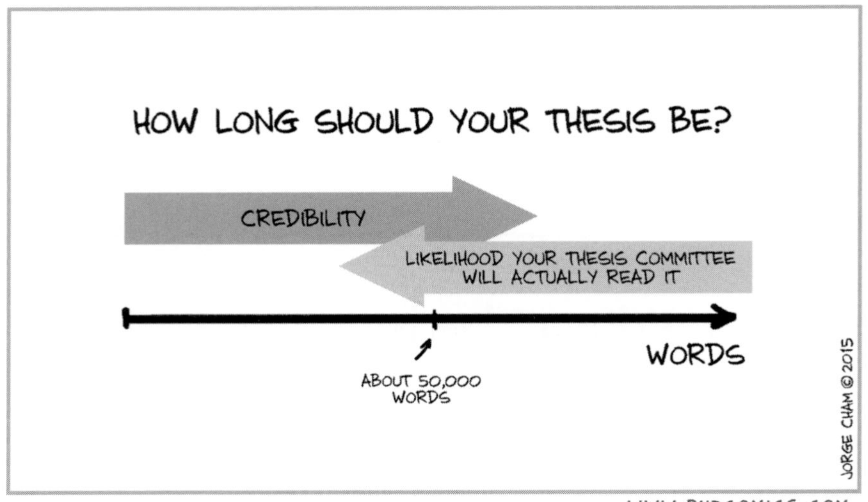

(www.phdcomics.com)

a candidate must produce an expansive traditional-style dissertation for a Ph.D. degree in the sciences must be abandoned. . . . The expansive traditional dissertation fosters the false impression that a typed record must be preserved of every table, graph, and successful or unsuccessful experimental procedure." Indeed, in some settings, the core of a thesis now normally consists of scientific papers that the student has published.

If a thesis serves any real purpose, that purpose might be to determine literacy. Perhaps universities have always worried about what would happen to their image if a Ph.D. degree turned out to have been awarded to an illiterate. Hence, the thesis requirement. Stated more positively, the candidate has been through a process of maturation, discipline, and scholarship. The "ticket out" is a satisfactory thesis.

It may be useful to mention that theses at European universities have tended to be taken much more seriously. They are designed to show that the candidate has reached maturity and can both do science and write science. Such theses may be submitted after some years of work and a number of primary publications, with the thesis itself being a "review paper" that brings it all together.

By the way, sometimes the word "dissertation" is used instead of "thesis." For example, at some U.S. universities, one speaks of a master's thesis but a doctoral dissertation. Whatever term one uses, the principles are much the same for preparing the less extensive master's-level document and the more extensive doctoral one.

TIPS ON WRITING

Few rules exist for writing a thesis, except those that may prevail in your own institution. Check whether your institution has a thesis manual or other set of instructions. If it does, obtain it and follow it carefully. Otherwise, your graduation may be delayed because of failure to use the required thesis format. If you do not have rules to follow, or even if you do, go to your departmental library—or, increasingly, your institutional library's online collection of theses—and examine the theses submitted by previous graduates of the department, especially those who have gone on to fame and fortune. Perhaps you will be able to detect a common flavor. Whatever ploys worked in the past for others are likely to work for you now.

Theses typically consist of several chapters. Sometimes the chapters correspond to the parts of an IMRAD scientific paper: introduction, methods, results, and discussion. Or, if a thesis reports a number of studies, the central part may include a chapter about each study. As noted, sometimes a thesis consists mainly of a set of published papers. In addition to chapters, common components of theses include a title page, acknowledgments, an abstract, a table of contents, a list of figures and tables, a list of abbreviations, and appendixes.

A thesis also should contain a substantial reference list, helping to demonstrate your familiarity with the literature in your field. In this regard, a thesis can resemble a review paper. Indeed, the introduction or a separate literature-review chapter generally should present a thorough review of previous work to which yours is related. Further, it is often desirable to go back into the history of your subject. You might thus compile a really valuable review of the literature of your field, while at the same time learning something about the history of science, which could turn out to be a most valuable part of your education.

Start with and work from carefully prepared outlines. Be careful about what goes in what section. If you have one or several results sections, the content must be your results, not a mixture of your results with those of others. If you need to present results of others, to show how they confirm or contrast with your own, you should do this within a discussion section. Otherwise, confusion may result, or, worse, you could be charged with lifting data from the published literature.

Give special attention to the introduction in your thesis for two reasons. First, for your own benefit, you need to clarify what problem you attacked, how and why you chose that problem, how you attacked it, and what you learned during your studies. The rest of the thesis should then flow easily and logically from the introduction. Second, first impressions are important, and you would not want to lose your readers in a cloud of obfuscation right at the outset.

Writing a thesis is a good chance to develop your skill in scientific writing. If a committee of faculty members is supervising your thesis research, seek

(www.phdcomics.com)

feedback on one or more drafts of your thesis from committee members, especially the committee chair. (If you can choose the committee members, try to include someone who writes very well and is willing to help others with writing.) Seek feedback early from the committee chair and others, to help prevent the need for extensive revisions at the end.

Universities keep copies of theses so that those interested can read them. Increasingly, they have been requiring electronic copies for this purpose. Use of electronic rather than paper copies saves space in libraries; can make theses easier for readers to obtain; and can let you include materials, such as videos or animations, that are difficult or impossible to provide in a bound thesis. Be sure to find out whether your institution requires you to submit your thesis electronically and, if so, what the instructions are.

WHEN TO WRITE THE THESIS

You would be wise to begin writing your thesis long before it is due. In fact, when a particular set of experiments or some major facet of your work has been completed, you should write it up while it is still fresh in your mind. If you save everything until the end, you may find that you have forgotten important details. Worse, you may find that you lack time to do a proper writing job. If you have not done much writing before, you might be amazed at what a painful and time-consuming process it is. You are likely to need a total of three months to write the thesis, on a relatively full-time basis. You will not have full time, however, nor can you count on the ready availability of your thesis advisor. Allow at least 6 months.

As implied in the preceding paragraph, a thesis need not be written from beginning to end. Work on the literature review section generally should start early, because your research should be based on previous research. Regardless of whether you do laboratory research or other investigation (such as field research or social-science surveys), you should draft descriptions of methods soon after methods are used, while memory is still complete. Often, the introduction is best drafted after the sections presenting and discussing the results, so the introduction can effectively prepare readers for what will follow.

Of course, ideas for any part of your thesis may occur to you at any time. Early in your research, consider establishing a physical or electronic file for each part of your thesis. Any time ideas occur to you for the thesis, put the ideas in the appropriate file. Similarly, if you come across readings that might be relevant to a given section, include or mention them in the file. Such files help keep you from losing ideas and materials that could contribute to your thesis. They also give you content to consider using as you begin drafting each section.

Perhaps you noticed that we said "drafting," not "writing." Much to the surprise of some graduate students, a good thesis is likely to require multiple drafts. Some graduate students think that once the last word leaves the keyboard, the thesis is ready to turn in. However, as with journal articles, considerable revision commonly is needed for the thesis to achieve its potential. Indeed, using feedback from one's graduate committee to strengthen the content, organization, and wording of one's thesis can be an important part of one's graduate education. Be prepared to need more time than expected to put your thesis in final form. Both in terms of the quality of the product and in terms of learning obtained that can aid in your future writing, the time is likely to be well spent.

RELATIONSHIP TO THE OUTSIDE WORLD

Remember, your thesis will bear only your name. Theses are normally copyrighted in the name of the author. The quality of your thesis and of any related publications in the primary literature probably will affect your early reputation and your job prospects. A tightly written, coherent thesis will get you off to a good start. An overblown encyclopedia of minutiae will do you no credit. The writers of good theses try hard to avoid the verbose, the tedious, and the trivial.

Be particularly careful in writing the abstract of your thesis. The abstracts of doctoral dissertations from many institutions are published in *Dissertation Abstracts International,* thus being made available to the larger scientific community.

Writing a thesis is not a hurdle to overcome before starting your scientific career. Rather, it is a beginning step in your career and a foundation for your later writing. Prepare your thesis carefully, and use the experience as a chance to refine your writing skills. The resulting document and abilities will then serve you well.

FROM THESIS TO PUBLICATION

People sometimes speak of "publishing a thesis." However, theses themselves are rarely, if ever, publishable. One reason is that theses commonly are intended partly to show that the graduate student has amassed considerable knowledge, and so they tend to contain much material that helps demonstrate scholarship but would not interest readers. Extracting one or more publications from a thesis generally entails considerable trimming and condensation. More specifically, writing one or more scientific papers based on a thesis requires determining what in a thesis is new and of interest to others and then presenting it in appropriate format and at an appropriate level of detail. In fields in which books present new scholarship, converting a dissertation to a book (Germano 2013) often includes decreasing the manuscript to a marketable length, dividing it into more chapters, using fewer quotations and examples, and otherwise making the manuscript more readable, cohesive, and engaging.

When you finish your thesis, promptly prepare and submit any manuscripts based on it, if you have not yet done so. Do so even if you are tired of your thesis topic—and tired from writing and defending a thesis. The longer you wait, the harder it is to return to your thesis and prepare a suitable manuscript based on it. And importantly, having one or more thesis-based writings published or in press can help catapult you into the next stage of your career.

CHAPTER 36

How to Prepare a Curriculum Vitae, Cover Letter, and Personal Statement

My one regret in life is that I am not someone else.

—Woody Allen

WHAT'S A CV? WHAT'S IT GOOD FOR?

Whereas those in business write résumés, we in the sciences generally prepare curricula vitae (CVs). Both a resume and a CV present key facts about one's professional background. However, the two differ somewhat in content and structure.

Literally, "curriculum vitae" means the course of one's life. A CV shows the course of your professional life. Figure 36.1 shows a CV of a fictional graduate student. Although the facts of this person's life are imaginary, the kinds of information provided are fairly typical: address and other contact information, education, honors, research, teaching, publications, and other professionally relevant experience.

A CV has many uses. You may be required to provide one with your thesis. Supplying one is standard when you apply for a job. Grant applications commonly include CVs. You will need to submit one if you are being considered for tenure, and you might need to provide one for your annual review. If you are nominated for an award, you may be asked to submit a CV to the selection committee. You should not, however, as one socially awkward young scientist did, offer your CV when asking someone for a date.

If you are seeking a position in industry, you may be asked for a résumé rather than a CV. Early in one's career, a CV and a résumé may be almost the same. However, a résumé commonly states an objective at the beginning. Also, duties generally are listed for jobs held. Whereas a CV can run several pages

Jian (Jane) Zhang

Address

123 College Street, #45
Anytown, TX 67890
(987) 654-3210
zhangjian@keepmail.com

Education

XYZ University, PhD, Molecular ABCology, expected May 2017
WXY University, MS, ABCology, August 2013
JKL College, BS, ABCology, *magna cum laude,* May 2011

Honors and Awards

Best Student Poster Presentation, International ABCology Society Annual Conference, October 2015
Regents Fellowship, XYZ University, 2013 to present
Outstanding Teaching Assistant Award, WXY University, 2013
Alumni Memorial Scholarship, JKL College, 2007-2011

Research

Molecular ABCology of DEF, laboratory of Sally Scientist, XYZ University, 2015–present
ABCological characteristics of GHI, laboratory of Rahm J. Researcher, WXY University, 2012–2013
Effect of D on E, National Institute of ABCology Summer Undergraduate Research Program, Washington, DC, June–August 2010

Teaching Experience

Instructor, Survey of ABCology, XYZ University, Fall 2015
Teaching Assistant, Introduction to ABCology, WXY University, 2012–2013
Tutor, JKL College Learning Center, 2009–2011 (ABCology, Calculus, Chinese)

Selected Extracurricular Activities

Graduate Student Association, XYZ University (member, 2013–present; co-organizer, Student Research Week, 2015–2016)
JKList (undergraduate research journal), JKL College (staff member, 2007–2009; editor, 2010–2011)

Publications

Scientist S, Smith JP, **Zhang J.** Inhibition of DEF by GHI. *Mol ABCol* (in press).
Zhang J, Researcher RJ. ABC chromatography of GHI. *ABCology* 2014;27:463–7.
Zhang J. ABCology: historical highlights and current trends. *JKList* 2010;5:79–81.

Figure 36.1. Curriculum vitae of a fictional graduate student.

or more, a résumé normally is limited to one or two pages, thus sometimes requiring that information be condensed. Many websites, books, and university career centers offer guidance on résumé preparation and provide sample résumés. If you need to prepare a résumé, consider using such resources.

WHAT TO PUT IN (AND WHAT TO LEAVE OUT)

Sometimes you may be told what types of information to include in a CV, what format to use, or both. For example, some colleges have detailed instructions for faculty CVs. Likewise, some funding agencies specify what to include in a CV in a grant application. Usual content and structure of CVs can differ among scientific fields and among institutions. Thus, it can be useful to look at others' CVs and have others review a draft of yours. For ideas of what to include in your CV and how to present it, consider looking online at the CVs of members of your department or of scientists elsewhere who are leaders in your field.

Do list your publications in your CV. Also list major presentations, such as papers given at national conferences. Consider listing as well the grants that you have received. In listing your publications, use a standard format for references (see Chapter 15), such as the one employed by a leading journal in your field. If a paper has been accepted but not yet published, list it as "in press" or "forthcoming." If it has been submitted but not yet accepted, or if it is still being prepared, do not list it under publications. You may, however, mention it in the research section of your CV.

Your CV should focus on your professional history. Normally, it should not include personal information such as date of birth, marital status, health, or hobbies. Do not list your Social Security number or other personal identification number, especially given the possibility of identity theft.

Of course, do not exaggerate your accomplishments. In addition to being dishonest, doing so can harm your career if the discrepancy is discovered. If there is nothing to list in a given category, omit that category. Do not be like the student who included the heading "Honors" in her CV and then wrote under it, "None."

OTHER SUGGESTIONS

CVs commonly are structured in reverse chronological order. In other words, within each category, items are listed from the most recent to the least recent. Some CVs, however, use chronological order. Whichever order is used, be consistent.

Do you use a nickname instead of your given name? If so, you may put it in parentheses. Ditto if you go by an English-language name in addition to the name in your native language. If readers might not be able to surmise your gender from your name and so might wonder how to address you, consider stating your gender in your CV or putting "Ms." or "Mr." in parentheses before your name. Of course, if you have a doctorate, those writing to you can simply use the gender-neutral "Dr."

Include some contact information that is unlikely to change, in case recipients wish to be in touch with you much later. For example, if you are a student, your address might well change. Therefore, consider also listing a long-term postal address (such as a parent's address) or including an email address or mobile phone number that is expected to stay the same.

If the nature of something listed might not be clear from its title, include a brief explanation in parentheses. You might say "Huth Award (for excellence in scientific writing)" or "Johnson Club (astronomy interest group)."

Consider having different versions of your CV for different uses. If you are seeking jobs at both research laboratories and teaching institutions, one version may focus mainly on your research experience and another may also list your teaching experience in detail. Even if the same information is included, it may appear in different orders in different versions of your CV.

Keep your CV up to date, so it is ready when needed. And, of course, proofread it carefully.

PREPARING A COVER LETTER

If you are applying for a job, you probably will need to accompany your CV with a cover letter. This letter provides further opportunity to introduce yourself, and it can help demonstrate your communication skills. Commonly, such a letter runs a single page. Rarely should it exceed two pages.

If possible, address the recipient of the letter by name. Be sure to spell the name properly. If it is unclear whether the recipient is a man or a woman, or whether the recipient has a doctorate, try to find out (for example, by checking online), so you can address the person appropriately. If this information is unavailable, address the person by full name (for example, "Dear Kelly Jones") rather than using a courtesy title (as in "Dear Mr. Jones"). If the name of the recipient is not available, you may use "To Whom It May Concern" or, if applicable, a more specific salutation such as "Dear Selection Committee." Do not use "Dear Sir" unless you are sure the recipient is male. In a formal letter, normally a colon rather than a comma follows the salutation.

At the beginning of the letter, make clear what you are applying for. Do not use general wording such as "the opening in your department," lest your application end up with those for the wrong position. Also consider including in the opening sentence your main qualifications. For example, you might say, "As a recent recipient of a Ph.D. in molecular ABCology from XYZ University, I am applying for the postdoctoral position in DEF research that was announced in *Science* last week."

In the middle of the cover letter, discuss your qualifications. You may introduce them by referring to your CV ("As noted in the accompanying curriculum vitae . . ."). Show how your qualifications match those requested in the position description. Doing so can provide a chance to elaborate on items listed in your CV. For example, you may summarize research you have done or identify techniques with which you are adept, or you may specify duties you had as a teaching assistant.

Do not discuss salary in your cover letter. Any such discussion should come later, once the employer expresses a desire to offer you a position.

End the letter positively but not overconfidently. Avoid overly assertive statements such as "Thus, I am the ideal candidate for the assistant professorship in molecular ABCology. I look forward to receiving an interview." An example of more appropriate wording: "Thus, I believe that my background qualifies me well for the assistant professorship in ABCology. I hope to hear from you soon about the possibility of an interview."

WRITING A PERSONAL STATEMENT

Applications for some opportunities require personal statements. For example, you may need to write a personal statement if you are applying to professional school or seeking some types of fellowships.

A personal statement is a brief essay that describes your professional development as it relates to the opportunity being sought. Often, it is best structured mainly in chronological order. You may begin with a paragraph providing a brief overview, then summarize how your interests have developed thus far, describe your main current activities, and finally discuss directions you anticipate taking. If feasible, show that your decision to seek the opportunity is well informed, for example by discussing related experience.

If you have a nontraditional background—for instance, if you pursued a different career before—or if you experienced a delay during your education, you generally should address the matter in your personal statement. Do not leave readers wondering why, for example, the dates in your CV do not seem to add up. If you discuss problems you have overcome, do so positively and without defensiveness, and show that you addressed the problems maturely and thoughtfully.

Be confident but not arrogant. In keeping with principles of good writing, show rather than tell. For example, to show that you have leadership abilities, you could state that you have held several leadership roles, note the main such roles, and mention a leadership award that you received. Do not emulate the medical-residency applicant who wrote, "First, I have a great bedside manner. . . . Second, I have excellent technical skills. . . . Third, and most importantly, I have a humble spirit."

In a personal statement, generally avoid or minimize discussion of aspects of your background that are not professionally related. In particular, do not discuss your political or religious views. Not only may doing so alienate readers whose views differ from yours; even if readers agree with your views, you may seem unprofessional or unfocused.

Finally, word your personal statement readably, in keeping with advice in this book. Those who review applications containing personal statements tend to be busy. Help them to understand quickly where you are coming from, where you are now, and where you are going. You will then be more likely to receive their support in obtaining the opportunity you seek.

CHAPTER 37

How to Prepare Grant Proposals and Progress Reports

The successful grant writer, to a large extent, lives by selling his ideas: a successful grant [proposal] is a marketing document.

—Janet S. Rasey

PREPARING A GRANT PROPOSAL

Scientific research costs money. Typically, the needed money comes as grants from government agencies, private foundations, or other sources. Thus, to survive professionally, most scientists must apply successfully for grants.

The purpose of a grant proposal, sometimes called a grant application, is to *persuade* a funding source to fund a project. To do so, it must *persuade* those making the decisions that

- the goal of the proposed research is worthwhile,
- the goal is relevant to the funding body's mission,
- the proposed research approach is sound,
- the staff is capable of doing the proposed work,
- adequate facilities will be available, and
- the requested amount of funding is reasonable.

Considerable competition exists for research funds, and careful preparation of a grant proposal can make the difference between being funded and not. As when writing a scientific paper, keys to success include using good models, following instructions carefully, and revising, revising, revising.

Identifying Potential Sources of Funding

How can you identify sources of funding that might be suitable for your work? During your research training, you probably became aware of major funding sources in your field. Indeed, if you were part of an active research group, your research supervisor might have spent considerable time writing grant proposals.

Your mentor may remain a good source of advice on finding funding sources. Colleagues and administrators also may be of help. At many institutions, grant offices and research offices publicize opportunities to apply for funding. Published or posted requests for proposals, and published or posted guides to funding opportunities, also can help. Email lists in your field or at your institution may include announcements of chances to seek funding, and Internet searching sometimes discloses further possibilities. Also, when you read scientific papers on work related to yours, notice the funding source, which may be specified in the acknowledgments section or in a note near the beginning or end of the paper. Doing so may disclose funding sources that you had not thought of pursuing.

As you identify potential sources of funding, start noticing their requirements for grant proposals. For example, when are the application deadlines? How does one access the instructions for applying? Can one proceed directly to submitting a grant proposal, or must one first submit preliminary information on what one wishes to propose?

Preliminary Letters and Proposals

Some funding sources require prospective grant applicants to begin by submitting preliminary information. Sometimes all that is required is a *letter of intent,* saying that one plans to respond to a given request for proposals and briefly describing the research that will be proposed. The funding source can then use this information to plan its work—for example, by starting to recruit peer reviewers with appropriate expertise to review your proposal once it arrives.

In other cases, prospective grant applicants must submit a *preliminary proposal,* sometimes also known by other names such as *letter of inquiry* or *preproposal.* A preliminary proposal is essentially a short version of the proposal one hopes to submit. On the basis of the preliminary proposals, the funding source, often with guidance from peer reviewers, decides which applicants can submit full proposals. Feedback about preliminary proposals can help applicants develop their full proposals and prepare future preliminary proposals.

Requiring preliminary proposals can save funding agencies the work of reviewing full proposals for research it is very unlikely to fund, and it can give them the opportunity to help shape research. It also can save scientists the work of preparing extensive proposals for research that the source is very unlikely to fund.

Preparing a preliminary proposal does, however, entail careful work. Much of the same rigorous thinking is required as for a full proposal. And writing concisely, so the preliminary proposal is informative despite being brief, can pose special challenges. (For guidance on communicating concisely, see Part VII of this book.) Because the opportunity to submit a full proposal is at stake, time spent writing and refining a preliminary proposal can be a valuable investment.

Common Parts of a Proposal

If your preliminary proposal is accepted, or if you will apply directly for a grant, determine the appropriate size and structure of the proposal. The instructions are likely to provide at least some guidance in this regard. Some aspects, however, may be left to your judgment.

Proposals range greatly in length, depending on the requirements of the funding source. Proposals for small internal grants at universities sometimes are limited to one page. Major proposals can run many pages.

Regardless of length, a good proposal generally includes background information relating to the proposed endeavor, a statement of goals, a research plan (or a program plan, in the case of an education or service project), a budget, and information about the qualifications of those who are to do the work (for example, curricula vitae). If a proposal runs several pages or more, it may well include a title page and an abstract.

Especially if a proposal is lengthy, other items may be required or advisable. These may include a letter of transmittal (analogous to the cover letter accompanying the manuscript for a scientific paper), a table of contents, a list of tables, a list of figures, a description of the predicted impact of the project, a plan for disseminating results, and information on facilities. A substantial research proposal generally cites references and includes a reference list.

Some proposals include appendixes for reviewers to consult if they want further information. Items that appendixes sometimes contain include scientific papers that have been accepted but not yet published, letters of support from prospective collaborators, and additional details about activities planned. Authors of grant proposals should keep in mind that reviewers typically are not obligated to look at appendixes. Thus, all key information should appear in the main body of the proposal.

Preparing to Write the Proposal

Only if a proposal matches the priorities of a funding source is it likely to be funded. Therefore, before writing, make sure the proposed work falls within the scope of what the source supports. Look carefully at written materials from the source in this regard. Also, feel free to consult staff members at the funding source (sometimes known as *program officers*) whose role includes advising prospective grant applicants. As well as saying whether a proposed project is likely to be considered seriously for funding, such individuals may be able to advise you on how to gear a proposal to help maximize chances of success. If one funding source seems to be a poor match, seek another.

In preparing grant proposals, as in other scientific writing, following good models saves time, avoids guesswork, and promotes success. If possible, look at one or more examples of successful proposals for the same category of grant from the same funding source. Colleagues who have received such grants may be willing to share copies of their proposals. Likewise, staff at the funding source may be able to provide examples. Other examples of well-prepared grant proposals, or of material therefrom, appear in books on grant application (for example, Gerin and Kapelewski 2011) or more broadly on technical writing (for example, Penrose and Katz 2010). In addition, examples of successful grant proposals have been posted on the web (for example, at www.niaid.nih.gov/research funding/grant/Pages/appsamples.aspx).

Writing the Proposal

Start working on the proposal long before the application deadline. For a lengthy proposal, at least 6 months beforehand can be advisable, particularly if others will collaborate in preparing it. Especially if you have little experience writing proposals or if written English is not among your strengths, consider obtaining help from a professional scientific writer or editor, either at your institution or on a freelance basis. For greatest effectiveness, such an individual should be involved early; when handed a proposal the day before it is due, an editor generally can do little more than make superficial improvements.

Read all instructions carefully, and follow them precisely. Be sure to provide all required information, and strictly follow requirements regarding length and other aspects of format. Realize, for example, that commonly a funding agency requires a biographical sketch in its own specialized format, rather than accepting a regular curriculum vitae; take the time to prepare or update your biosketch in keeping with current specifications. Proposals not complying with instructions may be disqualified without review. So, before submitting the proposal, check the instructions again.

Match the technical level of the proposal to the background of the reviewers. Government agencies typically have scientists in the researcher's field evaluate the grant proposal; thus, a proposal to such a funding source should be fairly technical. At some private funding sources, however, boards containing interested laypeople evaluate proposals. In the latter case, the proposal may need to be no more technical than a science article in a popular magazine. If in doubt as to how much background reviewers will have and therefore how technical the proposal should be, consult the funding agency.

Whatever the background of the reviewers, the proposal should be readably written. Scientists of sufficient prominence to review proposals are among the busiest in their fields, and commonly they have many proposals to review; they lack time to puzzle over what a proposal means, and so those proposals that

DREAMS OF ACADEMIC GLORY

"Dear Professor Hummel:
Not only have we awarded you a grant, but we have decided that your beautifully written grant application is publishable as it stands."

can be easily read and rapidly understood have an advantage. And of course, readable writing aids comprehension by lay reviewers of proposals. For readability, organize the writing carefully; present overviews before details; use simple, common language where possible; avoid wordy phrases; make effective (but not excessive) use of devices such as headings, boldface, and italics; and otherwise follow guidelines for readability. If doing so would aid communication, include tables, graphs, or other visuals in the proposal, if permitted. Of course, make sure that any such items are well prepared and suitably placed.

If a proposal is to include an abstract, devote particular care to it. An informative, well-organized, clearly worded abstract can be important for a number of reasons: Some funding sources choose reviewers for a proposal at least partly on the basis of the abstract; therefore, if an abstract is misleading or confusing, the proposal may be assigned to reviewers who are not the most suitable and thus it might not receive the most valid review. Also, reviewers generally gain their first impression of a proposal by reading the abstract, and so a poor abstract may bias the reviewers unfavorably. And reviewers commonly reread abstracts to refresh their memories before discussing proposals; at this stage too, a good abstract serves the applicant well.

Give the proposal a clear, concise title too. Doing so makes your focus apparent immediately, aids in capturing readers' attention, and helps create a good initial impression. Such a title also makes it easy for reviewers and others to refer to your proposal. Also, drafting a succinct, unambiguous title can help prepare you to write a strong, focused proposal (Friedland and Folt 2009). Do not feel obligated, though, to retain the initial title. As you and others prepare and refine the grant proposal, the title too may benefit from revision. The main point: The title deserves careful attention. It should not be a near-afterthought, added the hour before the proposal is due.

For many proposals, the applicant must use forms from the funding source. These forms commonly can be accessed through the web. Often, the completed forms constituting proposals can—or must—be submitted electronically. Carefully follow the instructions for preparation and submission.

If part or all of the proposal will consist of freestanding text, format it readably. If the funding source specifies items such as typeface, type size, and margins, be sure to follow the instructions. If such items are not specified, you generally should use a standard typeface (for example, Times Roman), 10- to 12-point type, and margins of 1 inch (about 25 mm) or slightly more. Also, unless otherwise stated, the right margin should be unjustified (ragged) rather than justified (straight). Do not use tiny type or minuscule margins in order to fit more words on the allotted pages; rather than helping your case, doing so is likely to rile the reviewers and thus undermine it.

Common Reasons for Rejection

Experienced reviewers of grant proposals have noted common reasons for rejection, as have staff members at funding agencies. By knowing and avoiding these problems, you can increase the likelihood that your proposal will be accepted.

A common reason for rejection—and presumably an easy one to avoid—is simply failure to follow the instructions for application. Poor writing, or otherwise sloppy presentation, also contributes to rejection. So does seeming unfamiliarity with relevant published work. Review the literature carefully, and cite it where appropriate; be sure all citations are accurate. Remember, scientists reviewing your proposal probably know thoroughly the literature in your field. Indeed, they may well have written items you cite or should be citing.

Other reasons for rejection include lack of originality, a superficial or unfocused research plan, and lack of a valid scientific rationale. (Are one or more well-conceived hypotheses being tested, or is the proposed research just a "fishing expedition," in hopes of finding something interesting?) Problems with the experimental approach—for instance, lack of suitable controls or failure to mention, if relevant, methods you plan to use if initial methods fail—also can lead to rejection. So can lack of experience with key methods (or failure to disclose such experience). And so can absence of enough experimental detail to persuade reviewers that the research is carefully planned. Looking at proposals accepted by the funding source can aid in determining how much detail to include and how technical the description of methods should be.

In many contexts, the word "ambitious" is a compliment. Not so, in general, regarding grant proposals. Proposing an unrealistically large amount of work can lead to rejection. Remember that experimental difficulties, unrelated interruptions, and other factors can slow a project. It is better to propose a somewhat modest endeavor that reviewers feel confident you can complete than one that appears too ambitious.

Unrealistic budgeting also can contribute to rejection. Carefully determine anticipated costs. If a budget is much too high, you may appear naive or greedy. You may likewise seem naive if the budget is much too low—and woe to you if the proposal is then approved and you are left to do the project with insufficient funds.

Other Problems to Watch For

Also take care to avoid other common problems—some substantive, some editorial.

For some types of research, proposals typically contain preliminary data. Find out whether such data are expected, and proceed accordingly.

Justify budgetary items sufficiently. Do not, for example, expect a funding source to cover the cost of a new computer or a trip to Hawaii unless you show why it is important to the proposed project. At many institutions, staff members who are experts on preparing grant budgets can provide assistance. Such an individual has published a chapter (Lewis 2008) giving researchers detailed guidance on grant-budget preparation; it includes a fictitious example of an extensive budget and budget justification.

If you are proposing a service project, for instance in science education, include sufficient information on plans for evaluating it. Especially for such projects, consider including a timeline to show that you have carefully planned what is to be accomplished when.

Edit the proposal carefully. In doing so, be especially alert for inconsistencies—which can arise if, for example, you alter the research plan but neglect to revise the abstract accordingly. Also be alert for confusingly extensive use of abbreviations. In general, use only or mainly those abbreviations that reviewers of the proposal should already know. If many abbreviations will be used, consider including a table of definitions for reviewers to consult.

Resubmitting a Proposal

If your proposal is not funded, do not be overly discouraged. Funding sources commonly receive proposals for many more projects than they can support. And some funding sources often accept revised versions of proposals they rejected on first submission.

Especially if the reviewers' feedback is favorable overall, try, try again, either by submitting a revised proposal to the same funding source or by seeking funding from another source. In preparing a revised proposal, as in revising a scientific paper for resubmission, make good use of suggestions from the reviewers.

If you are submitting a revised proposal to the same funding source, you generally should accompany it with a list showing, point by point, how the reviewers' advice was followed. (Of course, check the instructions for resubmissions.) If the reviewers identified a problem and you decided to correct it in a way other than that suggested, say what you did and why. Also, if appropriate, indicate the changes typographically, for example by using the Track Changes feature of Word. Seriously consider consulting the program officer responsible for the grant program to which you are applying. The program officer, who probably observed the peer review of your proposal, may have extra insights on how to strengthen your proposal and almost certainly knows well the re-submission process. Therefore, he or she may be able to guide you helpfully regarding both the content and the presentation of your revised proposal.

Keep trying, for writing successful grant proposals can require both skill and persistence. In the long run, the important thing is to obtain sufficient funding for your work. Along the way, preparation even of proposals not funded can bring you knowledge, ideas, and contacts that will ultimately contribute to your work.

Two Closing Comments

Two final thoughts on preparing grant proposals:

First, a suggestion: As you prepare and refine your grant proposal, envision yourself writing scientific papers about the completed research. Will you have all the needed information? If not, revise your research plan.

Second, a comment on wording: People sometimes speak of "writing a grant." However, the grant is the money—not the proposal or application. When colleagues say they are writing grants, one is tempted to respond, "While you have your checkbook out, please write a grant for me."

WRITING A PROGRESS REPORT

Some funding sources for grants, and some other supporters or supervisors of work in science, require progress reports at given intervals during projects. These reports help readers determine whether the work is progressing adequately and thus whether adjustments should be made in the plans, the funding level, or both. The prospect of preparing such reports can spur those doing the work to keep up. Writing such a report can aid in assessing one's own progress and, if advisable, adjusting one's approach. Also, such reports can be useful to draw on in drafting presentations and scientific papers.

Basic Structure

If the intended recipient of a progress report specifies a structure to use, of course use it; if forms (for example, regarding use of funds) are required, complete them as instructed. As when preparing a grant proposal, also follow any other instructions. If you have access to relevant examples of progress reports, consult them as models.

Commonly, progress reports contain three main sections: background information, a description of current status, and conclusions. Typically, the background

"OUR PROPOSAL DIDN'T GET THE GRANT, BUT THEY WANT US TO TEACH PROPOSAL WRITING."

(© ScienceCartoonsPlus.com)

section mainly summarizes the project plan. The section on current status presents achievements thus far, compares progress made with that anticipated, and describes any important problems encountered. The conclusions section can provide an overall assessment and describe and justify proposed modifications of the original plan.

Some Suggestions

Before writing a progress report, review the proposal (or other written plan) for the work. In general, structure the progress report similarly to the proposal. For example, if the proposal included sections on three subprojects, include a section on each in the progress report, and use the same headings as before.

Be specific in your report. For example, include relevant numbers, names, and dates. If appropriate, include tables and figures. To guide readers, consider using headings and other typographic devices.

Strive to sound positive, competent, and confident. However, do not hide problems. If you identify problems, say how they are being addressed.

If you write a series of progress reports on a project—for example, annual reports on work supported by a 5-year grant—put each in the same basic format. In addition to making the report easier to write, a consistent structure aids readers in comparing the content of successive reports. With a word-processing program, you can easily copy your previous report and update it to yield the current one. Remember, however, to make all needed changes.

Edit your progress report carefully. Double-check it for accuracy, and try to ensure that it is complete, clear, and concise. Your report can then both document your progress and serve as continuing evidence of your professionalism.

CHAPTER 38

How to Write a Recommendation Letter—and How to Ask for One

That's the news from Lake Wobegon, where all the women are strong, all the men are good looking, and all the children are above average.

—Garrison Keillor

Even as graduate assistants, those in the sciences often are asked to write recommendation letters for undergraduate students. Later, scientists also receive requests for recommendation letters for graduate students, postdoctoral fellows, and peers. Writing recommendation letters can consume much time. However, with a well-organized approach, you can efficiently write good recommendation letters, thus serving qualified candidates well while conserving your time. Likewise, with a well-considered approach, you can considerately and effectively obtain recommendation letters when you need them for yourself.

DECIDING WHETHER TO WRITE THE LETTER

A request to write a recommendation letter is just that: a request. Thus, you can decline. If you cannot honestly provide a favorable assessment, or if you cannot complete the letter by the deadline, promptly decline the request, so the requester can seek another recommender. When you cannot provide a favorable recommendation, a tactful statement such as "I think someone who knows you better could write a more convincing recommendation" may send the requester seeking a letter from someone else. If the requester persists, blunter wording may be needed.

If you know requesters fairly well and think they may be seeking opportunities poorly suited to them, consider meeting to discuss the decision. The requester may provide information that will change your mind and help you to write a more persuasive letter. Or you may find that the requester agrees with you but feels pressured to seek the opportunity. ("I'd rather do field research, but my family has always wanted me to become a physician" or "I thought I'd be letting you down if I didn't seek the summer fellowship.") With you as an ally, the requester can then better pursue his or her best interests.

Requesters cannot reasonably expect you to write recommendation letters immediately. If you tend to receive many requests for recommendation letters, consider letting it be known how much notice you generally need.

If there are people for whom you would be especially pleased to write letters, tell them. Doing so can relieve them of needless stress and help ensure that well-qualified candidates receive strong recommendations.

GATHERING THE INFORMATION

In preparing a recommendation letter as in writing a scientific paper, preliminaries include obtaining instructions, gathering materials, collecting data, and familiarizing yourself with examples.

As well as finding out when the recommendation is due and how to submit it, gather materials that are needed or would be useful. These may include a recommendation form to complete (if an electronic link to one has not been provided), a description of the opportunity or honor for which the candidate is being recommended, a résumé or curriculum vitae of the candidate, and examples of the candidate's work. They may also include items from your files, such as grade lists from your courses and previous letters you have written on the applicant's behalf. If the candidate is to have filled out part of a recommendation form, check that he or she has done so completely.

Norms regarding content and length of recommendation letters can differ among fields and cultures. Therefore, if you have not seen recommendation letters of the type you are to write, try to obtain some examples. Senior colleagues in your field may be able to show you some recommendation letters they have written, or they may offer feedback on a draft. If you serve on selection committees, you may see many such letters and gain a sense of the norm.

WRITING THE LETTER(S)

Having a usual format to follow can facilitate writing recommendation letters, just as it can aid in preparing a scientific paper. Here is one format that often works well:

In the first paragraph, indicate who is being recommended for what. An example of such a paragraph, which often runs only one sentence, is the following: "I am very pleased to recommend [name of candidate], a senior at [name of university], for admission to the graduate program in [name of field] at [name of university]." Placing the candidate's name in boldface can help recipients quickly see who is being recommended and file the letter appropriately.

In the next paragraph, say how you know the candidate. An example: "I have known Ms. [surname of applicant] for more than a year. As a junior, she took my course [title of course]. She also has worked in my laboratory since June through our university's undergraduate research program."

Then, in the following paragraph or two, provide your assessment of the candidate. Try to be specific. For example, rather than saying only that a candidate is an excellent student, specify the student's achievements, and perhaps rank the student relative to others. If applicable, note the candidate's academic or professional strengths and his or her relevant personal traits. Of course, gear what you say to what the person is being recommended for.

In the final paragraph, sum up. For example, you might write: "In sum, I consider Mr. [surname of applicant] an outstanding candidate for [name of opportunity]. I recommend him with enthusiasm." After a standard closing such as "Sincerely," Sincerely yours," or "Yours truly," sign your name. Your name and your professional title, such as assistant professor of [name of field], should appear under your signature. Normally, the letter should appear on official letterhead.

Sometimes a candidate may request several recommendations, for instance for graduate school or jobs. To be efficient, try to prepare all, or several, of the recommendations at once. Although, for example, different graduate programs may have different recommendation forms to complete, preparing the recommendations as a batch generally saves time. When there are forms, you may have the option of either writing your comments on them or attaching letters. If you already are writing a recommendation letter for a candidate, or if you are completing multiple recommendations for him or her, the latter option tends to be faster.

Especially if you think the candidate may later ask you to provide additional recommendations, keep copies of completed recommendation forms and save at least electronic files of recommendation letters. Preparing additional recommendations for the candidate should then be relatively quick and simple.

A LIGHT ASIDE

With regard to letters of recommendation, concern sometimes has existed that "the candidate may later exercise the legal right to read the letter, and perhaps even sue if the contents are not to his liking and are insufficiently substantiated." To address this concern, a professor at Lehigh University has devised a

"Lexicon of Inconspicuously Ambiguous Recommendations, or LIAR" (Thornton 1987). An example: "To describe a candidate who is not particularly industrious: 'In my opinion you will be very fortunate to get this person to work for you.'" Further examples along these lines appear in the book *L.I.A.R: The Lexicon of Intentionally Ambiguous Recommendations* (Thornton 2003).

IF YOU'RE SEEKING RECOMMENDATION LETTERS

What if you are seeking recommendation letters? The tips below—which follow largely from the advice above—can help you obtain them effectively and considerately.

Scientists and others providing recommendation letters generally are busy. Therefore, if possible, approach them well in advance. At a minimum, try to provide 2 weeks to write the letter. If you are asking for several recommendations, ideally provide at least 4 to 6 weeks.

If you think the potential recommender might not remember you at first, try to jog the person's memory. For example, if approaching the person by email, perhaps attach a photo of yourself. Or provide other identifying information, such as the topic on which you prepared a presentation.

Gauge the recipient's reaction to the request. If the person seems glad to write the recommendation, promptly provide the information needed to do so. But if he or she seems hesitant or is slow to reply, ask whether finding another recommender might be wise. You may save yourself from an awkward situation or a late or lukewarm recommendation.

Supply, in an organized way, items required to prepare the recommendation(s) well. Such items may include, in addition to needed forms, your curriculum vitae or résumé, descriptions of programs to which you are applying, and samples of your work.

Recommenders sometimes tell you, by email or otherwise, when the recommendations go out. If you do not hear, a polite inquiry a few days before the deadline can be appropriate.

Follow up on the recommendation. Thank the recommender, at least by email; especially if someone has written multiple recommendations, a thank-you card can be nice. When you gain your objective, inform the recommender. For instance, say where you will attend graduate school or embark on a job—and thank the recommender again.

In short, treat recommenders as you would wish to be treated in such roles. With luck, you will indeed be treated the same way.

CHAPTER 39

How to Work with the Media

When a reporter approaches, I generally find myself wishing for a martini.
—Jonas Salk

BEFORE THE INTERVIEW

Your scientific paper will be published soon, and a news release about it has attracted reporters. Or an earthquake, epidemic, or policy issue has drawn attention to your topic. Or maybe you are receiving an award. For whatever reason, a reporter calls. How can you work with the reporter to help ensure that the public receives accurate scientific information?

First, why work with the reporter? If your research is government funded, the public has a right to know. Also, as an important part of our culture, science merits coverage. Scientific information can help individuals and groups make sound decisions. Public information about science can draw students to scientific careers. Coverage in the popular media can promote public support for science and your institution.

At your institution, members of the media relations staff may prepare news releases and help reporters find experts to interview. They can also give guidance, such as tips on being interviewed for television. Other sources of advice include the SciDev.Net Practical Guide "What Journalists Want from Scientists and Why" (www.scidev.net/global/communication/practical-guide/what-journalists-want-from-scientists-and-why.html) and material from professional organizations such as the American Association for the Advancement of Science (see www.aaas.org/page/strategies-working-reporters) and the American Geophysical Union (see sharingscience.agu.org/inform-news). Books

"REMEMBER, A BREAKTHROUGH IS NOT A BREAKTHROUGH UNLESS YOU HAVE SOME GOOD P.R. FOR IT."

(© ScienceCartoonsPlus.com)

providing guidance include *A Scientist's Guide to Talking with the Media* (Hayes and Grossman 2006), *Communicating Science: A Primer for Working with the Media* (Menninger and Gropp 2008), *Am I Making Myself Clear? A Scientist's Guide to Talking to the Public* (Dean 2009), and *Explaining Research* (Meredith 2010).

News releases (also known as *press releases*) informing reporters about your research may be prepared by your institution or by the journal publishing your paper. They are then disseminated to the media. A news release, which can be published as is or can lead to a story by a reporter, is structured like a newspaper article; for many examples of news releases, see the science news website EurekAlert! (www.eurekalert.org). Those preparing a news release about your work will normally consult you. By answering their questions and then checking a draft, you can help ensure accuracy. Realize, though, that a news release will be much less technical and much less detailed than a scientific paper.

When reporters contact you, ask about their background, task, and timetable. Those writing science stories range from general reporters with minimal science background to science journalists with doctorates in the sciences; knowing whether the reporter is a specialist can help you respond appropriately. Also find out what the reporter is seeking; for example, will the article focus on your research, or is a general article being written about your research field? Finally, what is the reporter's deadline? Is the reporter writing a news story due today or a feature article due next month? Knowing the answers to such questions can help you respond most suitably. Of course, if you lack the expertise being sought, decline the interview and, if possible, direct the reporter to someone appropriately qualified.

Unless the reporter must talk with you immediately, think beforehand about what you want to say. Identify the main message you wish to present. Especially for the broadcast media, come up with a short and snappy way to state it—in other words, a "sound bite."

Before the interview, if possible, provide written materials or direct the reporter to some. Such materials may include news releases, papers you have written, and sources of general information about your research topic. Providing such materials facilitates the reporter's work, promotes efficient use of interview time, and fosters accuracy.

DURING THE INTERVIEW

When interviewed, try to word your responses in ways directly suitable for the reporter's audience. For example, use mainly simple common language, define technical terms, and relate what you say to familiar concepts, for instance by providing analogies. Consider presenting the information as you would to a nonscientist neighbor or a bright high school student. Suiting the material for the audience minimizes the need for the reporter to "translate" and so decreases the chance of error. It also gives the reporter quotable content or sound bites.

Try to present information accessibly but without condescension. Avoid thinking of "watering things down," which tends to yield indigestible bits in an insipid broth. Rather, think of "building bridges" between what you will present and what the audience already knows and cares about. Consider using techniques presented in Chapter 26, "How to Write for the Public," to present your content clearly and engagingly.

If you have key points to convey, make them even if the reporter does not ask. You may be able to do so by reframing a question. ("That's an interesting idea, but actually the issue we were studying was. . . .") Alternatively, you can add points at the end of an answer or the end of the interview. Also, if you have photographs or other visuals that might enhance the story, inform the reporter, even if not asked.

Stay focused during the interview. In particular, do not make offhand remarks that you would not want published.

Consider checking the reporter's understanding. For instance, you can say, "I'm not sure I've presented this concept clearly. Perhaps you could explain it

(© ScienceCartoonsPlus.com)

back to me so I can check." Then, if misunderstanding has occurred, you can provide clarification.

Before the interview ends, encourage the reporter to contact you if questions arise while writing the story. You may also offer to review part or all of the story for technical accuracy. Traditionally, journalists have not shown drafts to their sources, for fear of being pressured to change content inappropriately. Some journalists, however, welcome such review when writing about technical topics. Limit any suggested changes to matters of technical accuracy. The writing style is the writer's and editor's domain.

AFTER THE INTERVIEW

Once the story appears, have realistic expectations. Of course, the story will be briefer and less technical than a journal article. It also will focus mainly on aspects of greatest interest to the public. Thus, it is likely to emphasize conclusions and implications, and it is unlikely to describe your methods in detail or to list your eight coauthors. In evaluating the story, often the relevant question is not "Was everything captured precisely?" but rather "Would a member of the public come away with the correct idea?"

If a story is especially good or if it has serious errors, consider providing feedback. Often, reporters hear only if others are displeased. If you think the reporter has done especially well, tell the reporter—and, if possible, also tell the editor. If a story has a major inaccuracy, also inform the reporter. Good reporters want to know, so they can avoid repeating mistakes. If an error is serious, a correction may be published or aired.

Finally, think back on your interactions with the reporter. What did you do that turned out well? What could you have done better? Considering such questions can help you be even more effective the next time a reporter calls.

CHAPTER 40

How to Provide Peer Review

Honest criticism is hard to take, particularly from a relative, a friend, an acquaintance, or a stranger.

—Franklin P. Jones

RESPONDING TO A REQUEST FOR PEER REVIEW

Once you have become an author, you may receive invitations to be a peer reviewer—in other words, to evaluate work by others in your field. Journal editors may ask you to review papers being considered for publication. Funding agencies may ask you to review grant proposals. Book editors may ask you to review the proposals or manuscripts for books. Given that preparing peer reviews can entail much time and effort, why should you accept such invitations? And when should you decline them?

Peer review helps editors decide what to publish, and it helps authors improve their work. Similarly, peer review of grant proposals helps funding agencies make sound decisions and helps scientists refine their research. Others in your field provide this service to you when they review items you have written. Being a good citizen in the scientific community includes providing this service in return.

In addition, peer reviewing can have other benefits. It can help you keep up in your field and maintain your critical skills. Listing entities for which you peer review can enhance your curriculum vitae. Peer reviewing for a journal can lead to serving on its editorial board and becoming an editor of the journal. Although peer reviewing generally is unpaid, sometimes reviewers are paid or otherwise compensated. For example, reviewers of book manuscripts commonly receive a little money in appreciation of their efforts; if they prefer, sometimes they receive books from the publisher instead.

Sometimes, though, you should refuse the invitation to peer review or ask the editor whether to refuse it. If you lack time to complete the review adequately by the deadline, decline the opportunity and, if possible, suggest other potential reviewers. Also, if you believe you lack sufficient expertise to prepare a sound review, inform the editor. The editor may then ask you to suggest potential reviewers whom you consider well suited. Or the editor may explain that you were approached because of your expertise regarding an aspect of the research and that other reviewers will evaluate other aspects.

Inform the editor if you have conflicts of interest—that is, anything in your background that could interfere, or appear to interfere, with your objectivity in doing the review. For instance, if you have collaborated with any of the authors, if you have a financial interest relating to the research, or if an author is your friend or enemy or former spouse, tell the editor. Some journals routinely ask potential reviewers to state anything that might be a conflict of interest. Even if the journal does not, inform the editor if you think you might have one. The editor can then decide whether to retain you as a peer reviewer while keeping the item in mind or whether to seek a different reviewer.

PEER REVIEWING A SCIENTIFIC PAPER

If you are a peer reviewer, realize that the item being reviewed is confidential. Do not reveal its content. Do not discuss with those around you the authors' writing skills (or lack thereof). Do not ask others to collaborate on the review without first obtaining permission from the editor. If there is a valid reason for collaboration—for instance, if a colleague could better evaluate part of the research, or if collaborating on the review could help educate a graduate student or postdoctoral fellow—the editor is likely to grant permission. However, permission should be sought, not assumed.

Journals commonly use online systems for submission of peer reviews. Whether or not it does so, a journal is likely to seek two types of input from each reviewer: a confidential evaluation for use by the editors only, and comments for the editors to share with the authors. Some journals supply forms for these purposes. The form for feedback to editors may contain rating scales and provide room for comments about the quality of the work and whether the work should be published. Examples of items that the rating scales may address are the importance of the research question, the originality of the work, the validity of the methods used, the soundness of the conclusions, the clarity of the writing, and the suitability for the journal's readership.

Your comments for the editor to share with the authors typically should begin by saying what you perceive as the main strengths and main limitations of the paper. You should not, however, tell the authors whether you consider

"THAT'S IT? THAT'S PEER REVIEW?"

(© ScienceCartoonsPlus.com)

the paper publishable in the journal; that decision is up to the editor. After the general comments, you generally should provide a section-by-section list of comments on the paper. For ease in identification, it usually is best to specify the items you comment on by page, paragraph, and line.

Your main task as a peer reviewer is to evaluate the content of the paper. Is the research of high quality? If not, what are the problems? Has all the appropriate content been provided? Should any content be deleted? In answering the last two questions, you may find it useful to review the sections of this book on the respective sections of a scientific paper. Other potentially useful resources include a checklist (Task Force of *Academic Medicine* and the GEA-RIME Committee 2001) that appeared in a report providing guidance for peer reviewing. Although some items in this checklist apply only to some types of research, the checklist provides a useful framework.

As a peer reviewer, you are not expected to comment in detail on the writing. Your task does not include identifying every punctuation error and misspelling; if the paper is accepted, a copyeditor can correct such problems. However, it can be worthwhile to comment in general on the clarity, conciseness, and correctness of the writing; to note passages that are ambiguous; to suggest any reorganization that could improve the paper; and to remark on the design of figures and tables. If the paper contains highly specialized wording that you think a copy editor might have difficulty revising properly, consider providing some guidance. Also consider giving extra help with wording if the author's native language is not English.

In preparing comments intended for the authors, remember that the authors are human beings. Almost certainly, they care greatly about their work, are sensitive about it, and will be most receptive to feedback if it is given in a constructive tone. Therefore avoid sarcasm, and phrase your comments tactfully. Set a positive tone by first stating the strengths of the paper; then, after offering suggestions, perhaps end the review with words of encouragement. Although the section-by-section or line-by-line comments should be mainly suggestions, an occasional compliment can be included. Whether or not the journal accepts the paper, the review can help educate authors and so improve the current paper and later ones. Indeed, if an author appears to be a beginning researcher or seems to come from someplace where international norms of scientific publication are not well known, consider taking extra effort to make the review educational, either directly or by suggesting resources that can improve one's scientific writing.

Should you sign your review, or should it be anonymous? Policies in this regard differ among journals. Advocates of anonymous review, which is common in the sciences, say it allows reviewers to be more honest—especially when, for instance, a young researcher is evaluating a paper by someone much senior. On the other hand, advocates of signed review say it encourages reviewers to be more responsible. Some journals allow reviewers to decide whether to identify themselves. The journal's instructions for reviewers should indicate its policy. If in doubt, ask whoever invited you to do the review.

PROVIDING INFORMAL PEER REVIEW

Because of your knowledge of science and of writing, you may be asked to comment on drafts before submission. Such review can be a valuable service, espe-

(© Aries Systems Corporation)

cially to students, junior colleagues, and others who may not be thoroughly versed in English-language scientific communication. The following suggestions can aid in providing such informal peer review.

Find out what level of review is being sought. For example, is the draft an early one, and thus is the author seeking feedback mainly on content and organization? Or is the draft nearly final, so that the time has come to comment on fine points of expression? Although you should feel free to note problems on other levels, knowing the type of feedback sought can help you make appropriate use of your time.

Consider serving a criticism sandwich: praise, then criticism, then praise. Also show sensitivity to the author's feelings in other ways. For example, express criticisms as perceptions rather than facts ("I found this section hard to follow" rather than "This section is totally unclear"). And criticize the work, not the person ("This draft seems to contain many punctuation errors," not "You have a dreadful command of punctuation"). If you are providing feedback electronically on a manuscript, use word-processing features such as Track

Changes, or distinguish comments by placing them in triple brackets or using colored type, italics, or boldface. Avoid typing comments in all capital letters, which can give the impression that you are screaming. Similarly, if you write comments on hard copy, consider using green ink, which seems friendlier than red ink but also tends to be easy to notice.

Through providing informal feedback, you are teaching: By following your suggestions, authors can both improve their current drafts and become better writers in the long run. And by assimilating what you say and how you say it, they themselves can learn to be better peer reviewers in both the informal and the formal sense.

CHAPTER 41

How to Edit Your Own Work

There is no great writing, only great rewriting.

—Justice Louis Brandeis

PREPARING TO EDIT YOUR WORK

If you have reached this point in the book, or if you turned immediately to it, you probably know that good writing is much-revised writing. But how should you approach editing your own work? What should you look for? Is there anyone who can help? This chapter addresses these questions. In doing so, it reviews some key points from earlier chapters, for example about features of good scientific writing.

Challenges in editing one's own work include gaining distance and objectivity. Letting time elapse and changing the physical appearance of your work can help. If feasible, set your writing aside for at least a few hours. You might then be able to approach the piece much as a reader would. Perhaps also change the look of the piece (Hancock 2003) to aid in encountering the writing afresh. For instance, if you have been viewing the writing on a computer, print it out. Or change the typeface or margins. Maybe print the piece on colored paper. Such changes can assist in viewing your writing with new eyes.

Also use your ears. Read your draft aloud. In doing so, you may notice more easily where words are missing or wording is awkward.

Once you are ready to edit, in what order should you proceed? The choice is yours. Some authors start by considering large-scale aspects, such as overall content and organization. No need, they say, to bother with details right away, since parts of the writing might be deleted. Other authors start by polishing the language so they can see the piece more clearly before considering larger-scale changes. Such polishing can start with the text or with elements such as tables, figures, or references. Regardless of the order you use, thoroughly editing your work (or anyone else's) usually entails more than one round of editing. The final round should proceed from beginning to end, so you can better notice problems in the order of items.

ITEMS TO NOTICE: 8 Cs

Professional editors sometimes speak of checking for the "4 Cs"—which, depending on which editor you ask, can stand for "clarity, coherency, consistency, and correctness" (Einsohn 2011, p. 3) or variants such as "correctness, clarity, consistency, and courtesy." When editing your own work, consider checking for "8 Cs": compliance, completeness, composition, correctness, clarity, consistency, conciseness, and courtesy. Writing that achieves all eight Cs is likely to excel at communication, which is the C that is the overall goal.

Regarding *compliance,* ensure that the writing complies with all instructions, such as journals' instructions to authors and funders' directions for grant proposals. In addition, ensure that you have complied with relevant conventions in your field, for example regarding terminology and document structure. If your research involves animals or human subjects, also confirm that you have documented compliance with requirements in that regard.

Check for *completeness.* Does the document contain all necessary components? Does each component contain all the information that it should? Are necessary details included, for instance in the methods section?

Evaluate the *composition* of the piece. Is the overall structure appropriate? Is every section logically organized? Are paragraphs well structured, with strong topic sentences? Does one idea lead smoothly to the next? Are tables and figures well designed?

Check *correctness* of content and expression. Make sure all information is correct, both in the body of the text and in tables, figures, and references. See whether all the logic is valid. Also ensure that the grammar, spelling, punctuation, and word use are proper. If some aspects of such mechanics pose particular difficulty for you, devote special attention to them. For example, if you struggle with verb tenses, perhaps review your draft an extra time, checking specifically for them.

Pay attention to *clarity.* If some words or phrases might be unclear to readers, make sure they are defined. Likewise ensure that abbreviations are defined on first use. See whether antecedents of words such as "it" are clear. Look for places where the wording could be made clearer or where relationships of ideas could be clarified by using transitional words (such as "also," "first," "then," and "however"). Seek to improve passages where your reasoning might not be explicit enough for readers to follow. Identify sentences that are too long or too complex to understand easily, and divide or otherwise restructure them.

Look for *consistency* as well. Is all the information consistent—or, for example, do numbers differ between table and text? Is the content of the abstract consistent with that of the body of the piece? Is the terminology consistent throughout? Is the formatting, for example of subheadings, consistent? Do items appear in consistent order? Where appropriate, are tables and figures consistently formatted?

Both to save space and to aid readability, try to increase *conciseness*. In keeping with the examples in Appendix 2, replace long words with shorter equivalents and condense wordy phrases. Remove redundancies. Delete tangential or irrelevant content. In seeking conciseness, however, take care not to decrease clarity or diminish meaning.

Finally, keep *courtesy* in mind. Make sure you have been courteous to those whose work you discuss, others you mention, and your readers. Replace language that is unintentionally derogatory (such as "Previous researchers have failed to explore") with neutral language (such as "Previous researchers did not explore"). Revise any language that is not inclusive or that seems disrespectful of some population groups. To be courteous to readers, make sure you have attended to other Cs that help make writing easy to read—such as clarity, conciseness, and compliance with conventions. The effort that you invest in editing can save your readers effort and thus help ensure that your work will be read, understood, and appreciated.

A GOOD CHOICE: CHECKLISTS

Checking for all eight Cs at once can be hard or impossible. Therefore, generally check writing in several phases. To guide yourself, consider using checklists.

One good strategy is to use both a core checklist on general aspects of writing and a specialized checklist geared to the type of document that you have drafted. Figure 41.1 is an example of a core checklist. Figures 41.2 and 41.3 are examples, respectively, of specialized checklists for a scientific paper and for a grant proposal. Consider using these checklists or modifying them to suit your needs. Also consider obtaining or developing such checklists for other types of documents. Of course, if your target journal or other intended recipient provides a checklist, be sure to consult it.

Some General Questions to Ask: Editing One's Own Writing

1. Is the content complete, or should any content be added?
2. Should any content be deleted?
3. Is all the content accurate?
4. Is all the logic sound?
5. Do the content and crafting of the piece suit the audience?
6. Does the piece follow appropriate conventions regarding overall format?
7. If subheadings are allowed, are they used effectively?
8. Are sections and paragraphs of appropriate length?
9. Should any tables or figures be added or deleted?
10. If tables or figures are included, are they well designed?
11. Would typographic devices, such as italics of bullets, be helpful anywhere?
12. Is the piece well organized at various levels?
13. Are grammar, spelling, punctuation, and usage correct throughout?
14. Are verb tenses appropriate?
15. Are antecedents of all pronouns clear?
16. Have all acronyms been defined (and are all the acronyms worth using)?
17. Are sentences of appropriate length and structure?
18. If references are cited, are they in the appropriate format? Do all cited references appear in the reference list, and are all listed references cited in the text?
19. Is the writing clear, exact, and concise?
20. Have all instructions been followed?

Figure 41.1. Sample core checklist for editing one's own writing. A version of this checklist also appears in Gastel B. 2015. Editing and proofreading your own work. AMWA J. 30(4):147–151.

FINDING AND WORKING WITH AN AUTHOR'S EDITOR

Especially if you are a beginning author, consider seeking guidance from a manuscript editor. Individuals known as *author's editors* specialize in revising authors' work before submission. They can also help authors after submission, for example in improving a paper as requested by a journal.

How can you find an author's editor or the equivalent? Some universities, research institutions, and departments employ editors to assist scientists and scientists-in-training. In fact, some, such as the Mayo Clinic and The University of Texas MD Anderson Cancer Center, have scientific-publication units with multiple editors to provide such help. There also are freelance author's editors

Specialized Checklist: Editing One's Draft of a Scientific Paper

1. Does the title accurately and concisely indicate the content?
2. Are the appropriate people listed as authors?
3. Does the abstract accurately reflect the content of the paper? Is the abstract a suitable length?
4. Does the introduction provide sufficient context?
5. Does the introduction make clear what gap the research was intended to fill?
6. Does the introduction indicate the hypotheses or research questions?
7. Does the methods section provide sufficient information to replicate the research?
8. Does the methods section provide sufficient information to evaluate the research?
9. In the methods section, are sources of materials and equipment identified?
10. If the research was on humans or animals, are appropriate approvals noted?
11. Are the results presented in logical order?
12. Are the results presented in appropriate detail?
13. Are statistics appropriately presented?
14. Does the discussion address the hypotheses or research questions posed in the introduction?
15. Does the discussion put the results in sufficient context?
16. If relevant, does the discussion address strengths and weaknesses of the research?
17. If relevant, does the discussion identify applications or implications of the research?
18. Have the appropriate parties been acknowledged?

Figure 41.2. Example of a checklist for editing one's own scientific paper. Such a checklist could best be used along with a more general editorial checklist, such as shown in Figure 41.1. A version of this checklist also appears in Gastel B. 2015. Editing and proofreading your own work. AMWA J. 30(4):147–151.

and freestanding editorial companies. Networking with fellow researchers can aid in finding editors and editorial services that others regard highly. Lists of individuals who identify themselves as freelance editors appear on the websites of some universities, for example through the thesis office or writing center. Editors available for freelance work who have passed a rigorous examination in life science editing can be identified through the Board of Editors in the Life Sciences website, www.bels.org. Some commercial editing

SPECIALIZED CHECKLIST: EDITING A DRAFT OF ONE'S GRANT PROPOSAL

1. Does the title clearly and accurately convey the focus?
2. Is the abstract informative and clear? Ditto for any other sections serving as summaries?
3. Are the goals or hypotheses clear?
4. Is the originality of the work apparent?
5. Is the proposed work clearly relevant to the mission of the funding source?
6. Is the importance of the proposed work explained?
7. Is sufficient context provided?
8. Is the amount of proposed work realistic?
9. Is it clear that the personnel are capable of doing the proposed work?
10. Are sufficient justifications provided for choices, for example of methods?
11. Is sufficient supporting evidence included?
12. Is sufficient justification provided for budgetary items?
13. If there will be cost sharing, is sufficient information provided?
14. If preliminary studies are required or advisable, is there enough information about them?
15. If a timeline would be advisable, is one included?
16. If evaluation plans are needed, are they sufficient?
17. If dissemination plans should be included, are they sufficient?

Figure 41.3. Example of a checklist for editing one's own grant proposal. Such a checklist could best be used along with a more general editorial checklist, such as shown in Figure 41.1. A version of this checklist also appears in Gastel B. 2015. Editing and proofreading your own work. AMWA J. 30(4):147–151.

services are listed at www.authoraid.info/en/resources/details/750, and editorial guidance from volunteer mentors can be sought through the AuthorAID project (www.authoraid.info/en and www.authoraid.info/es). Although authors may benefit most from an editor who can meet with them face-to-face, email and other communication technologies allow effective use of an author's editor in another city or even another country.

Before giving your writing to an author's editor, edit it yourself insofar as feasible. Doing so helps use the editor's time efficiently, which may be especially desirable if the editor has many authors wanting help or if you yourself will pay the editor for the time spent. More important, doing some editing yourself can make the writing easier for the editor to understand, thus facilitating provision of suitable editorial feedback.

Communicate with the editor or editorial service about the desired extent of editing. Do you want the editor only to correct errors in grammar, punctuation, and other mechanics? Are you seeking more extensive help, including improvement of wording and sentence structure? Or do you desire whatever may strengthen the piece, including reorganization if deemed advisable? Ideally, once the editor has looked at the writing, find out what level of editing he or she considers suitable, and discuss how to proceed. Be available to answer questions. Realize that an author's editor is an advisor and thus that final decisions about the writing are yours.

A good author's editor, like a good peer reviewer, also serves as a good teacher. Notice revisions that the editor makes, and learn from them. If you are uncertain why a recurrent or major change was made, ask the reason if circumstances permit. Maybe compile a master list of changes made, to help avoid similar problems in future writing. If you use checklists in editing your work, perhaps revise them to reflect insights gained from the editor's feedback. In short, make the editor an ally and instructor in editing your own work.

CHAPTER 42

How to Seek a Scientific-Communication Career

Reporting on science and writing about it is like attending a never-ending graduate school of unlimited diversity.

—David Perlman

CAREER OPTIONS IN SCIENTIFIC COMMUNICATION

Some of us in science find the communication aspect so well suited to our interests and abilities that we focus on it in our careers. And increasingly, careers in science communication are being recognized as alternatives to those in research. Opportunities include writing and editing materials for fellow scientists and communicating science to general audiences.

Because you know your scientific discipline, its community, and its culture, you can bring much to publications for other scientists. Various niches exist in this realm. At a major journal, you may be an editor determining scientific content, a manuscript editor, or a writer or editor for the news section, if any. At a magazine or newsletter for scientists, you may be a writer or an editor. At a publisher of scientific books, you may be an acquisitions editor, generating topics, recruiting authors, and overseeing evaluation of proposals and manuscripts. At a corporation focusing on science or technology, you may write or edit. At a university or research institute or on a freelance basis, you may be an author's editor, working directly with authors to refine writing before submission. Because English is the international language of science, considerable demand exists for author's editors with strong English-language skills. An article by Kanel and Gastel (2008) summarizes career options in science editing.

Alternatively, you may pursue a career in the popular communication of science. For example, you may be a science reporter, writer, or editor for a newspaper, newsletter, magazine, or online publication. Or you may work in the broadcast media. You may write popular science books. You may prepare public-information materials for an organization or government agency concerned with science, technology, the environment, or medicine. At a university, you may write news releases, work on a research magazine, or pursue other public-communication activities. Likewise, you may work in media relations or public communication for a corporation. You may help prepare exhibits or other items for science museums. On a freelance basis, you may write about science for various media and institutions. Similarly, you may prepare science-communication materials for various outlets as an employee of a consulting firm.

AN ADMITTEDLY UNVALIDATED QUIZ

Many from the sciences who choose communication careers seem to share certain traits. Based on this observation, below is an informal quiz. As social scientists might point out, this quiz has not been validated systematically. But the smiles of recognition it evokes seem to suggest, literally, a sort of face validity.

So, is a science communication career for you? To help find out, consider the following 10 items:

1. Have you enjoyed courses in both science and other fields? Did you consider majoring in English or another area of liberal arts? Did you minor in such an area?
2. Are you an avid reader? Do you find yourself editing what you read? Do topics for writing often occur to you?
3. Do you like word games? For example, do you enjoy working crossword puzzles and playing Scrabble?
4. Have teachers or others complimented you on your writing?
5. In high school or college, did you serve on the school newspaper or another student publication? If not, did you consider doing so?
6. Do you consider yourself a science generalist? Rather than wanting to focus on a narrow research area, do you like to learn about various aspects of your field or of science in general? Do you find yourself more interested in knowing what other researchers are doing than in doing your own research?
7. Do you like to view science in its broad context? Are you interested not only in research itself but also in its applications and implications?
8. In laboratory projects, are you often the team member who writes things up? Do you find this role satisfying?

9. Do others ask you to edit what they wrote? Do they otherwise approach you for help with their writing?
10. Does a science communication career sound like fun to you? Is writing or editing something you would look forward to doing each day?

If you answered "yes" to most of these questions, a career in science communication might well be for you. And if on reading these questions you exclaim, "That's me!"—let us be the first to welcome you to the field.

CAREER PREPARATION

Some people enter scientific-communication positions directly from science. Serving as a peer reviewer and on the editorial board of a journal can lead to such a position. Some formal training, though, seems increasingly common, especially for those wishing to work in the popular communication of science. Such training can be in science journalism, scholarly publishing, technical communication, or a related field. It can consist of a degree program, a certificate program, or simply one or more courses. Resources for identifying educational opportunities include the Society of Environmental Journalists list of environmental journalism programs and courses (www.sej.org/library/education-environmental-journalism-programs-and-courses), which includes some listings more generally in science communication; an analogously broad listing (www.amwa.org/toolkit_new_med_writers#Related Resources) from the American Medical Writers Association; and the Society for Technical Communication Academic Database (www.stc.org/education/academic-database).

Some organizations offer workshops or other brief instruction that can help one develop professional skills in scientific communication. For example, the Council of Science Editors precedes its annual meeting with several concurrent short courses on aspects of science editing. Likewise, the annual conference of the American Medical Writers Association includes a wide array of 3-hour workshops, some of which are available as self-study modules (www.amwa.org/amwa_self_study).

Reading on one's own also can aid in preparing for a career in scientific communication. If you wish to enter scholarly scientific communication, works that may be useful include, in addition to the current book, guides to writing papers in specific fields of science (such as Ebel, Bliefert, and Russey 2004; Lang 2010; Sternberg and Sternberg 2010; and Zeiger 2000), style manuals commonly used in the sciences, and *The Copyeditor's Handbook* (Einsohn 2011). New editions of the style manuals, and of some of the other books, appear periodically—so be sure to obtain the most recent edition. Works that can assist those hoping to enter popular science communication include *A Field Guide for Science Writers* (Blum, Knudson, and Henig 2006), *Ideas into Words: Mastering the Craft of Science Writing* (Hancock 2003), *Health Writer's Handbook* (Gastel 2005), *The Science Writers' Handbook* (Writers of SciLance 2013), and *Handbook for Science Public Information Officers* (Shipman 2015).

Internships or fellowships in the communication of science can strengthen your skills, increase your visibility to potential employers, and aid in exploring career options. Sites of internship or fellowship programs in the communication of science have included international research centers (such as CERN and the International Centre for Theoretical Physics), U.S. government entities (such as Fermilab and the National Cancer Institute), journals (such as *Science* and *JAMA: The Journal of the American Medical Association*), magazines (such as *Science News* and *The Scientist)*, and other settings such as the public information offices of universities and of organizations. Also, since the 1970s, the American Association for the Advancement of Science has placed science graduate students at media sites each summer through its Mass Media Science and Engineering Fellow Program. In addition, sometimes communication offices without formal internship programs are willing to host interns on request; thus, if there is a setting where you might like to do an internship, take the initiative to ask.

ENTERING THE FIELD AND KEEPING UP

How can you find job opportunities in scientific communication? Look at position announcements posted on employment websites, published in journals, and disseminated by groups such as the Council of Science Editors, the National Association of Science Writers, and the American Medical Writers Association. Keep informed about job openings through social media. Whether or not job opportunities are announced, make yourself known to potential employers. Network, through organizations and otherwise.

Those in scientific communication, like those in scientific research, need to keep up with new developments. The never-ending graduate school of a science-communication career can aid in staying current with science, and further reading and listening can help fill the gaps. With regard to scientific communication, relevant organizations can aid in keeping up with trends, technologies, and issues; obtaining practical advice; and establishing or maintaining a network of others doing similar work. Examples of such organizations include those mentioned in the preceding paragraph and the European Association of Science Editors, the World Association of Medical Editors, the Association of Earth Science Editors, the Society of Environmental Journalists, the

Association of Health Care Journalists, and the Society for Technical Communication, as well as associations in more general communication fields. Read the publications of such associations, attend their conferences if you can, and take advantage of their social media engagement. And as your career develops, consider helping with their educational activities. You may one day be helping others from science who are entering scientific-communication careers.

APPENDIX 1

Selected Journal Title Word Abbreviations*

Word	Abbreviation	Word	Abbreviation
Abstracts	Abstr.	Archives	Arch.
Academy	Acad.	Archivio	Arch.
Acta	No abbrev.	Association	Assoc.
Advances	Adv.	Astronomical	Astron.
Agricultural	Agric.	Atomic	At.
American	Am.	Australian	Aust.
Anales	An.	Bacteriological	Bacteriol.
Analytical	Anal.	Bacteriology	Bacteriol.
Anatomical	Anat.	Bakteriologie	Bakteriol.
Annalen	Ann.	Berichte	Ber.
Annales	Ann.	Biochemical	Biochem.
Annals	Ann.	Biochimica	Biochim.
Annual	Annu.	Biological	Biol.
Anthropological	Anthropol.	Biologie	Biol.
Antibiotic	Antibiot.	Botanical	Bot.
Antimicrobial	Antimicrob.	Botanisches	Bot.
Applied	Appl.	Botany	Bot.
Arbeiten	Arb.	British	Br.
Archiv	Arch.	Bulletin	Bull.

*These abbreviations are written without the period in many journals.

Word	Abbreviation	Word	Abbreviation
Bureau	Bur.	Ethnology	Ethnol.
Canadian	Can.	European	Eur.
Cardiology	Cardiol.	Excerpta	No abbrev.
Cell	No abbrev.	Experimental	Exp.
Cellular	Cell.	Fauna	No abbrev.
Central	Cent.	Federal	Fed.
Chemical	Chem.	Federation	Fed.
Chemie	Chem.	Fish	No abbrev.
Chemistry	Chem.	Fisheries	Fish.
Chemotherapy	Chemother.	Flora	No abbrev.
Chimie	Chim.	Folia	No abbrev.
Clinical	Clin.	Food	No abbrev.
Commonwealth	Commw.	Forest	For.
Comptes	C.	Forschung	Forsch.
Conference	Conf.	Fortschritte	Fortschr.
Contributions	Contrib.	Freshwater	No abbrev.
Current	Curr.	Gazette	Gaz.
Dairy	No abbrev.	General	Gen.
Dental	Dent.	Genetics	Genet.
Developmental	Dev.	Geographical	Geogr.
Diseases	Dis.	Geological	Geol.
Drug	No abbrev.	Geologische	Geol.
Ecology	Ecol.	Gesellschaft	Ges.
Economics	Econ.	Helvetica	Helv.
Edition	Ed.	History	Hist.
Electric	Electr.	Immunity	Immun.
Electrical	Electr.	Immunology	Immunol.
Engineering	Eng.	Industrial	Ind.
Entomologia	Entomol.	Institute	Inst.
Entomologica	Entomol.	Internal	Intern.
Entomological	Entomol.	International	Int.
Environmental	Environ.	Jahrbuch	Jahrb.
Ergebnisse	Ergeb.	Jahresberichte	Jahresber.

Word	Abbreviation	Word	Abbreviation
Japan, Japanese	Jpn.	Publications	Publ.
Journal	J.	Quarterly	Q.
Laboratory	Lab.	Rendus	R.
Magazine	Mag.	Report	Rep.
Material	Mater.	Research	Res.
Mathematics	Math.	Review	Rev.
Mechanical	Mech.	Revue, Revista	Rev.
Medical	Med.	Rivista	Riv.
Medicine	Med.	Royal	R.
Methods	No abbrev.	Scandinavian	Scand.
Microbiological	Microbiol.	Science	Sci.
Microbiology	Microbiol.	Scientific	Sci.
Monographs	Monogr.	Series	Ser.
Monthly	Mon.	Service	Serv.
Morphology	Morphol.	Society	Soc.
National	Natl.	Special	Spec.
Natural, Nature	Nat.	Station	Stn.
Neurology	Neurol.	Studies	Stud.
Nuclear	Nucl.	Surgery	Surg.
Nutrition	Nutr.	Survey	Surv.
Obstetrical	Obstet.	Symposia	Symp.
Official	Off.	Symposium	Symp.
Organic	Org.	Systematic	Syst.
Paleontology	Paleontol.	Technical	Tech.
Pathology	Pathol.	Technik	Tech.
Pharmacology	Pharmacol.	Technology	Technol.
Philosophical	Philos.	Therapeutics	Ther.
Physical	Phys.	Transactions	Trans.
Physik	Phys.	Tropical	Trop.
Physiology	Physiol.	United States	U.S.
Pollution	Pollut.	University	Univ.
Proceedings	Proc.	Untersuchung	Unters.
Psychological	Psychol.	Urological	Urol.

Word	Abbreviation	Word	Abbreviation
Verhandlungen	Verh.	Zeitschrift	Z.
Veterinary	Vet.	Zentralblatt	Zentralbl.
Virology	Virol.	Zoologie	Zool.
Vitamin	Vitam.	Zoology	Zool.
Wissenschaft-liche	Wiss.		

APPENDIX 2

Words and Expressions to Avoid

Sermons on brevity and chastity are about equally effective. Verbal promiscuity flows from poverty of language and obesity of thought, and from an unseemly haste to reach print—a premature ejaculation, as it were.

—Eli Chernin

Jargon	Preferred Usage
a considerable amount of	much
a considerable number of	many
a decreased amount of	less
a decreased number of	fewer
a great deal of	much
a majority of	most
a number of	many, some
a small number of	a few
absolutely essential	essential
accounted for by the fact	because
adjacent to	near, next to
along the lines of	like
an adequate amount of	enough
an example of this is the fact that	for example

Jargon	Preferred Usage
an order of magnitude faster	10 times as fast
apprise	inform
are of the same opinion	agree
as a consequence of	because
as a matter of fact	in fact (or leave out)
as a result of	because
as is the case	as happens
as of this date	today
as to	about (or leave out)
at a rapid rate	rapidly, fast
at an earlier date	previously
at an early date	soon
at no time	never
at present	now
at some future time	later, sometime
at the conclusion of	after
at the present time	now
at this point in time	now
based on the fact that	because
because of the fact that	because
by means of	by, with
causal factor	cause
cognizant of	aware of
completely full	full
consensus of opinion	consensus
considerable amount of	much
contingent upon	dependent on
count the number of	count
definitely proved	proved
despite the fact that	although
due to the fact that	because
during the course of	during, while
during the time that	while

Jargon	Preferred Usage
effectuate	cause
elucidate	explain
employ	use
enclosed herewith	enclosed
end result	result
endeavor (v.)	try
entirely eliminate	eliminate
eventuate	happen
fabricate	make
facilitate	help
fatal outcome	death
fellow colleague	colleague
fewer in number	fewer
finalize	end
first of all	first
following	after
for the purpose of	for
for the reason that	because
from the point of view of	for
future plans	plans
give an account of	describe
give rise to	cause
has been engaged in a study of	has studied
has the capability of	can
has the potential to	can, may
have the appearance of	look like, resemble
having regard to	about
immune serum	antiserum
impact (v.)	affect
implement (v.)	start, put into action
important essentials	essentials
in a number of cases	sometimes
in a position to	able to

Jargon	**Preferred Usage**
in a satisfactory manner	satisfactorily
in a situation in which	when
in a very real sense	in a sense (or leave out)
in almost all instances	nearly always
in case	if
in close proximity to	close, near
in connection with	about, concerning
in the light of the fact that	because
in many cases	often
in most cases	usually
in my opinion it is not an unjustifiable assumption that	I think
in only a small number of cases	rarely
in order to	to
in relation to	toward, to
in respect to	about
in some cases	sometimes
in terms of	about
in the absence of	without
in the event that	if
in the most effective manner	most effectively
in the not-too-distant future	soon
in the possession of	has, have
in this day and age	today
in view of the fact that	because
inasmuch as	for, as
inclined to the view	think
initiate	begin, start
is defined as	is
is desirous of	wants
is detrimental to	harms
is similar to	resembles

Jargon	Preferred Usage
it has been reported by Smith	Smith reported
it has long been known that	I haven't bothered to look up the reference
it is apparent that	apparently, clearly
it is believed that	I think (or say who thinks)
it is clear that	clearly
it is clear that much additional work will be required before a complete understanding	I don't understand it
it is evident that *a* produced *b*	*a* produced *b*
it is generally believed	many think
it is my understanding that	I understand that
it is of interest to note that	(leave out)
it is often the case that	often
it is suggested that	I think
it is worth pointing out in this context that	note that
it may be that	I think, perhaps
it may, however, be noted that	but
it should be noted that	note that (or leave out)
it was observed in the course of the experiments that	we observed
join together	join
lacked the ability to	could not
large in size	large
larger compared to	larger than
let me make one thing perfectly clear	a snow job is coming
majority of	most
make reference to	refer to
many different types	many types
met with	met
militate against	prohibit
more often than not	usually

Jargon	Preferred Usage
needless to say	(leave out, and consider leaving out what follows it)
new initiatives	initiatives
no later than	by
of an efficient nature	efficient
of great theoretical and practical importance	useful
of long standing	old
of the opinion that	think that
on a daily basis	daily
on account of	because
on behalf of	for
on no occasion	never
on the basis of	by
on the grounds that	because
on the part of	by, among, for
on those occasions in which	when
our attention has been called to the fact that	we belatedly discovered
owing to the fact that	because
perform	do
place a major emphasis on	stress, emphasize
pooled together	pooled
presents a picture similar to	resembles
previous to	before
prior to	before
protein determinations were performed	proteins were determined
quantify	measure
quite a large quantity of	much
quite unique	unique
rather interesting	interesting
red in color	red
referred to as	called

Jargon	**Preferred Usage**
regardless of the fact that	even though
relative to	about
resultant effect	result
root cause	cause
serious crisis	crisis
should it prove the case that	if
smaller in size	smaller
so as to	to
subject matter	subject
subsequent to	after
sufficient	enough
take into consideration	consider
terminate	end
the fact of the matter is that	(leave out)
the field of chemistry	chemistry
the great majority of	most, almost all
the opinion is advanced that	I think
the predominate number of	most
the question as to whether	whether
the reason is because	because
the vast majority of	most, almost all
there is reason to believe	I think
they are the investigators who	they
this particular finding	this finding
this result would seem to indicate	the result indicates
through the use of	by, with
to the fullest possible extent	fully
transpire	happen
ultimate	last
unanimity of opinion	agreement
until such time	until
utilization	use

Jargon	Preferred Usage
utilize	use
very unique	unique
was of the opinion that	believed
ways and means	ways, means (not both)
we have insufficient knowledge	we do not know
we wish to thank	we thank
what is the explanation of	why
whether or not to	whether to
with a view to	to
with reference to	about (or leave out)
with regard to	concerning, about (or leave out)
with respect to	about
with the exception of	except
with the result that	so that
within the realm of possibility	possible

APPENDIX 3

SI (Système International) Prefixes and Their Abbreviations

No.	*Prefix*	*Abbreviation*
10^{-18}	atto	a
10^{-15}	femto	f
10^{-12}	pico	p
10^{-9}	nano	n
10^{-6}	micro	μ
10^{-3}	milli	m
10^{-2}	centi	c
10^{-1}	deci	d
10	deka	da
10^{2}	hecto	h
10^{3}	kilo	k
10^{6}	mega	M
10^{9}	giga	G
10^{12}	tera	T
10^{15}	peta	P
10^{18}	exa	E

APPENDIX 4

Some Helpful Websites

Academic Phrasebank (www.phrasebank.manchester.ac.uk)
Provides lists of phrases to consider using in various parts of scientific papers.

Annotated Journal Article (www.authoraid.info/en/resources/details/648/)
A journal article with comments appearing in boxes on it. Illustrates many points in this book. An exercise that consists of writing such comments on an article in one's field may increase one's skill in scientific writing.

AuthorAID (www.authoraid.info/en/ and www.authoraid.info/es/)
A project mainly to help researchers in developing countries to write about and publish their work. Includes a resource library containing materials in several languages. Also includes a blog and provides opportunity to seek mentors. Has links to many other resources.

Beall's List (scholarlyoa.com/publishers/)
A list of "potential, possible, or probable predatory scholarly open-access publishers."

Board of Editors in the Life Sciences (www.bels.org)
Organization certifying editors in the life sciences through a rigorous examination. Website includes list of editors who are certified and thus may be well suited to edit manuscripts before journal submission.

Creating Effective Poster Presentations (www.ncsu.edu/project/posters/index.html)
Guidance on many aspects of poster presentation.

Creative Commons (creativecommons.org)
Source of free licenses that can serve as an alternative to transferring copyright. Using standardized licenses, authors can specify which uses of their work they permit, and under what conditions.

Designing Conference Posters (colinpurrington.com/tips/poster-design)
Extensive guidance on preparing poster presentations. Includes templates for designing posters.

The Elements of Style (www.bartleby.com/141/)
First edition of a classic book on basics of English-language writing.

Grammar Girl (www.quickanddirtytips.com/grammar-girl)
Advice on grammar, punctuation, word choice, and related topics.

How to Recognize Plagiarism (www.indiana.edu/~academy/firstPrinciples/choice.html)
Tutorials to help users understand and thus avoid plagiarism.

Instructions to Authors in the Health Sciences (mulford.utoledo.edu/instr)
Provides links to instructions to authors for thousands of biomedical journals. Also has links to related guidelines.

On Being a Scientist: A Guide to Responsible Conduct in Research, 3rd edition (www.nap.edu/catalog/12192/on-being-a-scientist-a-guide-to-responsible-conduct-in)
A guide to ethics in science from the (U.S.) National Academies. Includes material on ethics in scientific publication. Site provides access to the full text of the guide and to an accompanying video.

OneLook Dictionary Search (www.onelook.com)
Provides opportunity to seek definitions and related information in multiple dictionaries.

ORCID (orcid.org)
Initiative providing researchers with unique, persistent personal identification numbers (ORCID identifiers) that they can use to identify themselves unambiguously as the author of their scientific papers and other communications.

Phony vs Legit (allenpress.com/frontmatter/issue/issue-29-2014/phony-vs-legit)
Infographic to aid in distinguishing predatory journals from valid journals.

Recommendations for the Conduct, Reporting, Editing, and Publication of Scholarly Work in Medical Journals (www.icmje.org/icmje-recommendations.pdf)
Set of recommendations followed by many medical journals. Originally focused on format but now largely emphasizes ethical and other issues. Known in previous versions as the Uniform Requirements for Manuscripts Submitted to Biomedical Journals.

SciDev.Net Practical Guides (www.scidev.net/global/content/practical-guides.html)
Expert advice mainly on aspects of communicating science.

Glossary

Abstract. Brief synopsis of a paper, usually summarizing each major section of the paper. Different from a summary, which is usually a summary of conclusions. Also, a brief synopsis of a grant proposal.

Acknowledgments. The section of a paper (following the discussion but preceding references) designed to thank individuals and organizations for the help, advice, materials, or financial assistance they provided during the research and writing of the paper.

Acquisitions editor. An editor responsible for obtaining book manuscripts.

Address. Identifies the author and supplies the author's mailing address.

Ad hoc reviewer. *See* **Referee.**

Alphabet-number system. A system of literature citation in which references are arranged alphabetically in the references or literature cited, numbered, and then cited by number in the text. A variation of the name and year system.

Archival journal. This term is equivalent to "primary journal" and refers to a journal that publishes original research results.

Article-level metrics. Statistics regarding use of individual articles. Examples: numbers of downloads, mentions in social media, and citations.

Author. A person who actively contributed to the design and execution of the research and who takes intellectual responsibility for the research results being reported.

Author's editor. An editor who helps authors to improve manuscripts, proposals, or other documents before submission. May work for a research institution, for an editing company, or on a freelance basis.

Blog. Short for "weblog." An ongoing series of postings on a website. Typically written in first person

Book review. An article describing and evaluating a book.

Caption. *See* **Legend**.

Citation-order system. A system of referencing in which references are cited in numerical order as they appear in the text. Thus, the references section is in citation order, not in alphabetical order.

Compositor. One who sets type. Equivalent terms are "typesetter" and "keyboarder."

Conference report. A paper written for presentation at a conference. Most conference reports do not meet the definition of valid publication. A well-written conference report can and should be short; experimental detail and literature citation should be kept to a minimum.

Conflict of interest. In science, a situation in which financial or other personal considerations may interfere with a researcher's objectivity in conducting or reporting research.

Contributor. Someone who contributed to the research reported in a scientific paper or to the writing of the paper. May or may not qualify for listing as an author.

Copy editor. *See* **Manuscript editor**.

Copyright. The exclusive legal right to reproduce, publish, and sell written intellectual property.

Corresponding author. In a multiauthor paper, the author designated to receive and respond to inquiries from the journal editorial office and from readers.

Council of Science Editors. An organization whose members are involved mainly with the editing and publishing of journals in the sciences. Formerly the Council of Biology Editors. www.CouncilScienceEditors.org.

Creative Commons. An organization providing free licenses that state conditions under which specified works such as journal articles can be reproduced or otherwise used. creativecommons.org.

Cropping. The marking of a photograph so as to indicate parts that need not appear in the published photograph, or the electronic removal of material at the edges of a photograph. As a result, the essential material is "enlarged" and highlighted.

CSE. *See* **Council of Science Editors**.

Curriculum vitae. A document listing information about one's education and career. Commonly known as a "CV."

CV. *See* **Curriculum vitae**.

Deputy editor. The editor second in command at a publication with multiple editors.

Digital Object Identifier (DOI). An identification code, assigned to an online article, that provides a persistent link to its location on the Internet.

Digital poster. *See* **Electronic poster**.

Discussion. The final section of an IMRAD paper. Its purpose is to fit the results from the current study into the preexisting fabric of knowledge. The important points are expressed as conclusions.

Dual publication. Publication of the same data two (or more) times in primary journals. A violation of scientific ethics unless permission is obtained from the initial publication site and the republished material is clearly identified as such.

Editor. The title usually given to the person who decides what will (and will not) be published in a journal or in a multiauthor book.

Editorial. A brief article presenting opinion.

Editorial consultant. *See* **Referee.**

Editor in chief. The top editor of a publication with multiple editors. In charge of overall content.

Electronic poster (e-poster). Poster that is provided digitally and displayed electronically. Also known as a "digital poster."

Embargo. A policy of some journals stating that research reported in articles accepted by the journal cannot be reported elsewhere, such as in the popular media, before it appears in the journal.

Fabrication. Inventing research findings rather than obtaining them through scientific research. Clearly a major ethical violation.

Festschrift. A volume of writings by different authors presented as a tribute or memorial to a particular individual.

Graph. Lines, bars, or other pictorial representations of data. Graphs are useful for showing the trends and directions of data. If exact values must be listed, a table is usually superior.

Hackneyed expression. An overused, stale, or trite expression.

Halftone. A photoengraving made from an image photographed through a screen and then etched so that the details of the image are reproduced in dots.

Hard copy. When an old-fashioned manuscript on paper is provided via a word processor or computer, it is called hard copy.

Harvard system. *See* **Name and year system.**

Impact factor. A measure of the average number of citations per article published in a given journal, as determined by Journal Citation Reports. Sometimes used to indicate the relative prominence of a journal within a given discipline.

IMRAD. An acronym derived from introduction, methods, results, and discussion, the organizational scheme of most modern scientific papers.

Introduction. The first section of an IMRAD paper. Its purpose is to state clearly the problem investigated and to provide the reader with background information.

Jargon. *Merriam-Webster's Collegiate Dictionary*, 11th ed., defines jargon as "a confused unintelligible language."

Keyboarder. *See* **Compositor.**

Legend. The title or name given to an illustration, along with explanatory information about the illustration. Also called a "caption."

Letter of inquiry. *See* **Preliminary proposal.**

Letter of intent. Letter indicating to a funding source that one plans to submit a grant proposal.

Letter to the editor. A letter intended for publication in a journal or elsewhere.

Literature cited. The heading used by many journals to list references cited in an article. The headings "References" and (rarely) "Bibliography" are also used.

Managing editor. A title often given to the person who manages the business affairs of a journal. Typically, the managing editor is not involved with editing (acceptance of manuscripts). However, he or she may be responsible for copyediting (part of the production process).

Manuscript editor. A person (either an employee of the publisher or a freelance contractor) whose responsibility is to prepare manuscripts for publication by improving mechanics such as spelling and grammar, ensuring consistency with the required style, and providing markup for the typesetter or printer. Also known as a "copy editor."

Markup for the typesetter. Marks and symbols used by manuscript editors and sometimes authors to transmit type specifications to the typesetter or printer.

Masthead statement. A statement by the publisher, usually on the title page of the journal, giving ownership of the journal and a succinct statement describing the purpose and scope of the journal.

Materials and methods. *See* **Methods.**

Methods. The second section of an IMRAD paper. Its purpose is to describe the experiment in such detail that a competent colleague could repeat the experiment and obtain the same or equivalent results.

Monograph. A specialized, detailed book written by specialists for other specialists.

Name and year system. A system of referencing in which a reference is cited in the text by the last name of the author and the year of publication; for example, Smith (1990). Also known as the "Harvard system."

News release. A written announcement for journalists, for example regarding publication of a journal article. Structured much like a newspaper story. Also known as a "press release."

Offprints. *See* **Reprints.**

Open access. Refers to scientific papers (or other writings) that are available free of charge on the Internet to all who are interested.

Oral report. Similar in organization to a published paper, except that it lacks experimental detail and extensive literature citation. And, of course, it is spoken, not printed.

ORCID. Short for "Open Researcher and Contributor ID." An initiative that provides researchers with persistent, unique identification numbers, known as ORCID identifiers, used mainly to definitively identify journal authors.

Peer review. Evaluation of a manuscript by peers of the author (scientists working in the same area of specialization).

Plagiarism. Presentation of someone else's words and ideas as one's own rather than crediting the source.

Poster. In science, a display board presenting research. Also refers to the digital equivalent. *See* **Electronic poster**.

Predatory conference. Entity that the organizer promotes as being a valid conference but that instead is a ruse for taking money from prospective attendees.

Predatory journal. Entity that claims to be a legitimate journal but instead exploits authors by taking their money without providing valid publication.

Preliminary proposal. Brief initial proposal submitted to a funding source, which then determines whether it wishes to receive a full proposal. Also known by other terms, such as "letter of inquiry" and "pre-proposal."

Pre-proposal. *See* **Preliminary proposal**.

Press release. *See* **News release**.

Primary journal. A journal that publishes original research results.

Primary publication. The first publication of original research results, in a form whereby peers of the author can repeat the experiments and test the conclusions, and in a journal or other source document readily available within the scientific community.

Printer. Historically, a device that prints or a person who prints. Often, however, "printer" is used to mean the printing company and is used as shorthand for all of the occupations involved in the printing process.

Production editor. An editor who coordinates the editing of a book manuscript and other aspects of book production.

Program officer. Person managing part or all of the grant program for a funding source. Role can include advising grant applicants.

Proof. A copy of typeset material sent to authors, editors, or managing editors for correction of typographical errors.

Proofreaders' marks. A set of marks and symbols used to instruct the compositor regarding errors on proofs.

Publisher. A person or organization handling the business activities concerned with publishing a book or journal.

Query. A question a manuscript editor (copy editor) asks an author, for example about something in a manuscript that is unclear or inconsistent.

Query letter. A letter proposing a magazine article.

Referee. A person, usually a peer of the author, asked to examine a manuscript and advise the editor regarding publication. The term "reviewer" is used more frequently but perhaps with less exactness. Also sometimes called an "editorial consultant."

Reprints. Separately printed journal articles supplied to authors, usually for a fee. Sometimes called "offprints." Can be electronic.

Results. The third section of an IMRAD paper. Its purpose is to present the new information gained in the study being reported.

Reviewer. *See* **Referee.**

Review paper. A paper written to summarize and integrate previously published knowledge about a topic. Can be either an overview of a field or a critical, interpretive study of literature in the field. Also known as a "review article."

Running head. A headline repeated on consecutive pages of a book or journal. The titles of articles in journals are often shortened and used as running heads. Also called "running headlines."

Science writing. A type of writing whose purpose is to communicate scientific knowledge to a wide audience including (usually) both scientists and nonscientists.

Scientific editor. An editor, trained as a scientist, whose role is primarily to oversee evaluation of submitted papers and participate in deciding which ones to publish.

Scientific paper. A written and published report describing original research results.

Scientific writing. A type of writing whose purpose is to communicate new scientific findings to other scientists. Also sometimes includes other scientist-to-scientist communications, such as review articles and grant proposals.

SHERPA/RoMEO. Database of journal publishers' policies on copyright and self-archiving. www.sherpa.ac.uk/romeo.

Society for Scholarly Publishing. An organization of scholars, editors, publishers, librarians, printers, booksellers, and others engaged in scholarly publishing. www.sspnet.org.

Summary. Usually a summary of conclusions, placed at the end of a paper. Different from an abstract, which usually summarizes all major parts of a paper and appears at the beginning of the paper (heading abstract).

Syntax. The order of words within phrases, clauses, and sentences.

Systematic review article. Review article (review paper) based on use of systematic, explicit methods to gather and analyze literature on a well-defined question.

Table. Presentation of (usually) numbers in columnar form. Tables are used when many determinations need be presented and the exact numbers have importance. If only "the shape of the data" is important, a graph is usually preferable.

Thesis. A manuscript demanded of an advanced-degree candidate; its purpose is to prove that the candidate is capable of doing original research and writing about it. The term "dissertation" is essentially equivalent but should be reserved for a manuscript submitted for a doctorate.

Title. The fewest possible words that adequately describe the contents of a paper, book, poster, or so forth.

Typesetter. *See* **Compositor.**

References

Aaronson S. 1977. Style in scientific writing. In E Garfield (ed.), Essays of an information scientist, vol. 3, 4–13. Available at www.garfield.library. upenn .edu/essays/v3p004y1977-78.pdf. Accessed March 3, 2011.

American National Standards Institute. 1969. American national standard for the abbreviation of titles of periodicals. ANSI Z39.5-1969. New York: American National Standards Institute.

American National Standards Institute. 1977. American national standard for bibliographic references. ANSI Z39.29-1977. New York: American National Standards Institute.

American National Standards Institute. 1979a. American national standard for the preparation of scientific papers for written or oral presentation. ANSI Z39.16-1979. New York: American National Standards Institute.

American National Standards Institute. 1979b. American national standard for writing abstracts. ANSI Z39.14-1979. New York: American National Standards Institute.

Anderson JA, Thistle MW. 1947. On writing scientific papers. Bull Can J Res, December 31, 1947, NRC no. 1691.

Becker HS. 1986. Writing for social scientists: how to start and finish your thesis, book, or article. Chicago: University of Chicago Press.

Beer D, McMurrey D. 2014. A guide to writing as an engineer. 4th ed. Hoboken, NJ: Wiley.

Bernstein TM. 1965. The careful writer: a modern guide to English usage. New York: Atheneum.

Bishop CT. 1984. How to edit a scientific journal. Baltimore: Williams and Wilkins.

Blakeslee A. 1994. Late night thoughts about science writing. Quill, November/ December, 35–38.

Blum D, Knudson M, Henig RM, eds. 2006. A field guide for science writers. 2nd ed. New York: Oxford University Press.

Booth V. 1981. Writing a scientific paper and speaking at scientific meetings. 5th ed. London: The Biochemical Society.

Bornmann L, Mutz R. 2015. Growth rates of modern science: a bibliometric analysis based on the number of publications and cited references. J Assoc Inf Sci Technol. 66:2215–2222.

Briscoe MH. 1996. Preparing scientific illustrations: a guide to posters, presentations, and publications. 2nd ed. New York: Springer-Verlag.

CBE Style Manual Committee. 1983. CBE style manual: a guide for authors, editors, and publishers in the biological sciences. 5th ed. Bethesda, MD: Council of Biology Editors.

Chase S. 1954. Power of words. New York: Harcourt, Brace.

The Chicago manual of style. 2010. 16th ed. Chicago: University of Chicago Press.

Clark C. 2005. Conflict of interest and scientific publication: a synopsis of the CSE retreat. Science Editor. 28:39–43.

Claxton LJ. 2005. Scientific authorship. Part 2. History, recurring issues, practices, and guidelines. Mutat Res. 589:31–45.

Coghill AM, Garson LR, eds. 2006. The ACS style guide: effective communication of scientific information. 3rd ed. Washington, DC: American Chemical Society; New York: Oxford University Press.

Committee on Science, Engineering, and Public Policy. 2009. On being a scientist: a guide to responsible conduct in research. 3rd ed. Washington, DC: National Academies Press. Available at www.nap.edu/catalog/12192/on-being-a-scientist-a-guide-to-responsible-conduct-in. Accessed January 17, 2016.

Council of Biology Editors. 1968. Proposed definition of a primary publication. Newsletter, Council of Biology Editors, November 1968, 1–2.

Cromey DW. 2010. Avoiding twisted pixels: ethical guidelines for the appropriate use and manipulation of scientific digital images. Sci Eng Ethics. 16:639–667.

Cromey DW. 2012. Digital images are data: and should be treated as such. In DJ Taatges and J Roth (eds.), Cell imaging techniques: methods and protocols, 2nd ed., 1–27. New York: Humana Press.

Daugherty G. 1999. You can write for magazines. Cincinnati: Writer's Digest Books.

Davidoff F. 2000. Who's the author? Problems with biomedical authorship, and some possible solutions. Science Editor. 23:111–119.

Day RA. 1975. How to write a scientific paper. ASM News. 42:486–494.

Day RA. 1979. How to write and publish a scientific paper. Philadelphia: ISI Press.

Day RA, Sakaduski ND. 2011. Scientific English: a guide for scientists and other professionals. 3rd ed. Santa Barbara, CA: ABC-CLIO.

Dean C. 2009. Am I making myself clear? A scientist's guide to talking to the public. Cambridge, MA: Harvard University Press.

Ebel HF, Bliefert C, Russey WE. 2004. The art of scientific writing: from student reports to professional publications in chemistry and related fields. 2nd ed. Weinheim, Germany: Wiley-VCH.

Einsohn E. 2011. The copyeditor's handbook: a guide for book publishing and corporate communications. 3rd ed. Berkeley: University of California Press.

Fisher RA. 1938. Presidential address. Sankhyā: The Indian Journal of Statistics. 4(1):14–17. Available at http://www.jstor.org/stable/40383882. Accessed August 8, 2015.

Fowler HW. 1965. A dictionary of modern English usage. 2nd ed. London: Oxford University Press.

Frick T, Dagli C, Barrett A, Myers R, Kwon K. 2016. How to recognize plagiarism: tutorial and tests. Bloomington, IN: Department of Instructional Systems Technology, School of Education, Indiana University. www.indiana.edu/~academy/firstPrinciples/. Accessed January 17, 2016.

Friedland AJ, Folt CL. 2009. Writing successful science proposals. 2nd ed. New Haven, CT: Yale University Press.

Gahran A. 2000. Instant information. Writer's Digest, June 2000, 48–49.

Gahran A. 2001. Repurposing content for the Web. Writer's Digest, June 2001, 50–51.

Garfield E. 1999. Journal impact factor: a brief review. CMAJ. 161:979–980.

Gastel B. 1983. Presenting science to the public. Philadelphia: ISI Press.

Gastel B. 1991. A strategy for reviewing books for journals. BioScience. 41: 635–637.

Gastel B. 2005. Health writer's handbook. 2nd ed. Ames, IA: Blackwell.

Gerin W, Kapelewski CH. 2011. Writing the NIH grant proposal: a step-by-step guide. 2nd ed. Los Angeles: Sage.

Germano W. 2013. From dissertation to book. 2nd ed. Chicago: University of Chicago Press.

Godlee F, Jefferson T, eds. 2003. Peer review in health sciences. 2nd ed. London: BMJ Books.

Hall GM, ed. 2013. How to write a paper. 5th ed. Chichester, West Sussex, UK: Wiley-Blackwell.

Halm EA, Landon BE. 2007. Everything you wanted to know about writing a research abstract but were too afraid (or started too late) to ask. SGIM Forum 30(12):2, 13. Available at www.sgim.org/userfiles/file/Forum200712.pdf. Accessed March 3, 2011.

Hancock E. 2003. Ideas into words: mastering the craft of science writing. Baltimore: Johns Hopkins University Press.

Hartley J. 2007. Colonic titles! The Write Stuff. 16:147–149.

Hayes R, Grossman D. 2006. A scientist's guide to talking with the media: practical advice from the Union of Concerned Scientists. New Brunswick, NJ: Rutgers University Press.

Houghton B. 1975. Scientific periodicals: their historical development, characteristics and control. Hamden, CT: Shoe String Press.

Huth EJ. 1987. Medical style and format: an international manual for authors, editors, and publishers. Baltimore: Williams and Wilkins.

Huth EJ. 1999. Writing and publishing in medicine. 3rd ed. Baltimore: Williams and Wilkins.

International Committee of Medical Journal Editors. 2014. Recommendations for the conduct, reporting, editing, and publication of scholarly work in medical journals. www.icmje.org/icmje-recommendations.pdf. Accessed August 18, 2015.

Iverson C. 2002. US medical journal editors' attitudes toward submissions from other countries. Science Editor. 25:75–78. Available at www.councilscience editors.org/files/scienceeditor/v25n3p075-078.pdf. Accessed March 3, 2011.

Iverson C, Christiansen S, Flanagin A, Fontanarosa PB, Glass RM, Gregoline B, Lurie SJ, Meyer HS, Winker MA, Young RK. 2007. AMA manual of style: a guide for authors and editors. 10th ed. New York: Oxford University Press.

Kanel S, Gastel B. 2008. Careers in science editing: an overview to use or share. Science Editor. 31:18–22. Available at www.councilscienceeditors.org/wp -content/uploads/v31n1p018-022.pdf. Accessed January 17, 2016.

Keegan DA, Bannister SL. 2003. Effect of colour coordination of attire with poster presentation on poster popularity. CMAJ. 169:1291–1292.

Lang TA. 2010. How to write, publish, and present in the health sciences: a guide for clinicians and laboratory researchers. Philadelphia: American College of Physicians.

Lederer R. 1987. Anguished English. New York: Dell.

Lewis MA. 2008. The financing and cost accounting of science: budgets and budget administration. In LM Scheier and WL Dewey (eds.), The complete writing guide to NIH behavioral science grants, 347–395. New York: Oxford University Press.

Lock S. 1985. A difficult balance: editorial peer review in medicine. London: The Nuffield Provincial Hospitals Trust.

Maggio R. 1997. Talking about people: a guide to fair and accurate language. Phoenix: Oryx Press.

McGirr CJ. 1973. Guidelines for abstracting. Tech Commun. 25(2):2–5.

Menninger H, Gropp R. 2008. Communicating science: a primer for working with the media. Washington, DC: American Institute of Biological Sciences.

Meredith D. 2010. Explaining research: how to reach key audiences and advance your work. New York: Oxford University Press.

Meyer RE. 1977. Reports full of "gobbledygook." J Irreproducible Results. 22(4):12.
Michaelson HB. 1990. How to write and publish engineering papers and reports. 3rd ed. Phoenix: Oryx Press.
Mitchell JH. 1968. Writing for professional and technical journals. New York: Wiley.
Mitrany D. 2005. Creating effective poster presentations: the editor's role. Science Editor. 28:114–116. Available at http://www.councilscienceeditors.org/wp-content/uploads/v28n4p114-116.pdf. Accessed January 15, 2016.
Moher D, Liberati A, Tetzlaff J, Altman DG, and the PRISMA Group. 2009. Preferred reporting items for systematic reviews and meta-analyses: the PRISMA statement. Ann Intern Med. 151:264–269.
Morgan P. 1986. An insider's guide for medical authors and editors. Philadelphia: ISI Press.
Morrison JA. 1980. Scientists and the scientific literature. Scholarly Publishing. 11:157–167.
National Institutes of Health. 2010. Research integrity. Available at grants.nih.gov/grants/research_integrity/definitions.htm. Accessed January 27, 2016.
O'Connor M. 1978. Standardisation of bibliographical reference systems. Br Med J. 1(6104):31–32.
O'Connor M. 1991. Writing successfully in science. London: HarperCollins Academic.
O'Connor M, Woodford FP. 1975. Writing scientific papers in English: an ELSE-Ciba Foundation guide for authors. Amsterdam: Associated Scientific.
Outing S. 2001. Think locally, write globally. Writer's Digest, July 2001, 52–53.
Peat J, Elliott E, Baur L, Keena V. 2002. Scientific writing: easy when you know how. London: BMJ Books.
Penrose AM, Katz SB. 2010. Writing in the sciences: exploring conventions of scientific discourse. 3rd ed. New York: Longman.
Publication manual of the American Psychological Association. 6th ed. 2010. Washington, DC: American Psychological Association.
Ratnoff OD. 1981. How to read a paper. In KS Warren (ed.), Coping with the biomedical literature, 95–101. New York: Praeger.
Reid WM. 1978. Will the future generations of biologists write a dissertation? BioScience. 28:651–654.
Roig M. 2003. Avoiding plagiarism, self-plagiarism, and other questionable writing practices: a guide to ethical writing. Available at ori.hhs.gov/avoiding-plagiarism-self-plagiarism-and-other-questionable-writing-practices-guide-ethical-writing. Accessed August 8, 2015.
Rosner JL. 1990. Reflections on science as a product. Nature. 345:108.
Rowan KE. 1990. Strategies for explaining complex science news. Journalism Educator. 45(2):25–31.

Saleh N. 2013. The complete guide to article writing. Cincinnati: Writer's Digest Books.

San Francisco Declaration on Research Assessment. 2012. Available at www.ascb.org/dora/. Accessed January 17, 2016.

Schwartz M, and the Task Force on Bias-Free Usage of the Association of American University Presses. 1995. Guidelines for bias-free writing. Bloomington: Indiana University Press.

Scott-Lichter D, and the Editorial Policy Committee, Council of Science Editors. 2012. CSE's white paper on promoting integrity in scientific journal publications, 2012 update. Wheat Ridge, CO: Council of Science Editors. Available at www.councilscienceeditors.org/resource-library/editorial-policies/white-paper-on-publication-ethics. Accessed August 18, 2015.

Shipman M. 2015. Selfish reasons for researchers to publicize their study findings. Available at www.scilogs.com/communication_breakdown/selfish-reasons. Accessed August 22, 2015.

Shipman WM. 2015. Handbook for science public information officers. Chicago: University of Chicago Press.

Sternberg RJ, Sternberg K. 2010. The psychologist's companion: a guide to writing scientific papers for students and researchers. 5th ed. New York: Cambridge University Press.

Stocking SH, and the Writers of the New York Times. 2011. The New York Times reader: science and technology. Washington, DC: CQ Press.

Strunk W Jr, White EB. 2000. The elements of style. 4th ed. Boston: Allyn and Bacon.

Style Manual Subcommittee, Council of Science Editors. 2014. Scientific style and format: the CSE manual for authors, editors, and publishers. 8th ed. Chicago, IL: University of Chicago Press.

Sun XL, Zhou J. 2002. English versions of Chinese authors' names in biomedical journals: observations and recommendations. Science Editor. 25:3–4. Available at www.councilscienceeditors.org/wp-content/uploads/v25n1p003-004.pdf. Accessed January 17, 2016.

Tananbaum G. 2013. Article-level metrics: a SPARC primer. Available at www.sparc.arl.org/sites/default/files/sparc-alm-primer.pdf. Accessed August 9, 2015.

Task Force of Academic Medicine and the GEA-RIME Committee. 2001. Checklist of review criteria. Acad. Med. 76:958–959. Available at journals.lww.com/academicmedicine/Fulltext/2001/09000/APPENDIX_1__CHECKLIST_OF_REVIEW_CRITERIA.37.aspx. Accessed January 17, 2016.

Taylor RB. 2011. Medical writing: a guide for clinicians, educators, and researchers. 2nd ed. New York: Springer.

Thornton RJ. 1987. "I can't recommend the candidate too highly": an ambiguous lexicon for job recommendations. Chronicle of Higher Education, February 25, 42.

Thornton RJ. 2003. L.I.A.R: the lexicon of intentionally ambiguous recommendations. 2nd ed. Napierville, IL: Sourcebooks.
Trelease SF. 1958. How to write scientific and technical papers. Baltimore: Williams and Wilkins.
Tuchman BW. 1980. The book: a lecture sponsored by the Center for the Book in the Library of Congress and the Authors League of America. Washington, DC: Library of Congress.
Vence T. 2015. Know your PIO. The Scientist. 29(1):67–69.
Waquet F. 2008. Posters and poster sessions: a history. American Institute of Physics History Newsletter. Available at repository.aip.org/islandora/object/nbla:275060#page/5/mode/2up. Accessed January 17, 2016.
Ware M, Mabe M. 2015. The STM report: an overview of scientific and scholarly journal publishing. 4th ed. The Hague, Netherlands: International Association of Scientific, Technical and Medical Publishers. Available at www.stm-assoc.org/2015_02_20_STM_Report_2015.pdf. Accessed August 8, 2015.
Weiss EH. 1982. The writing system for engineers and scientists. Englewood Cliffs, NJ: Prentice-Hall.
Weiss EH. 2005. The elements of international English style: a guide to writing correspondence, reports, technical documents, and Internet pages for a global audience. Armonk, NY: Sharpe.
Wiley S. 2009. Bring back reprint requests. The Scientist. 23(9):29.
Writers of SciLance. 2013. The science writers' handbook. Boston: Da Capo Press.
Zeiger M. 2000. Essentials of writing biomedical research papers. 2nd ed. New York: McGraw-Hill.
Zerubavel E. 1999. The clockwork muse: a practical guide to writing theses, dissertations, and books. Cambridge, MA: Harvard University Press.

Index